# Electrostatic Hazards

# Electrostatic Hazards

Dipl.-Ing. Günter Lüttgens
and Dr Norman Wilson

BUTTERWORTH
HEINEMANN

Butterworth-Heinemann
Linacre House, Jordan Hill, Oxford OX2 8DP
A division of Reed Educational and Professional Publishing Ltd

A member of the Reed Elsevier plc group

OXFORD BOSTON JOHANNESBURG
MELBOURNE NEW DELHI SINGAPORE

First published 1997



**British Library Cataloguing in Publication Data**
A catalogue record for this book is available from the British Library

ISBN 0 7506 2782 4

**Library of Congress Cataloguing in Publication Data**
A catalogue record for this book is available from the Library of Congress

Composition by AccComputing, Castle Cary, Somerset
Printed in Great Britain by Biddles of Guildford & Kings Lynn

# Contents

# Foreword

The topic of electrostatic hazards in industry has long been recognized as an important but often misunderstood subject. On the one hand, too often accidents have occurred because of a lack of understanding of the fundamental principles of the origin of static electricity. On the other, in the course of accident investigations when no obvious cause could be identified, electrostatics has often been blamed inappropriately.

This book fills a real need for industrial practitioners by providing insight to the nature of static electricity and specific examples of problems that can arise in the workplace. Although aimed at the non-specialist the book contains much new information never previously published which will be of interest to the specialist as well. The authors bring combined experience of over 60 years which they have devoted to the subject and have managed to summarize here.

The material is divided into three main sections. Chapters 1–3 include basic background information dealing with the dangers of fire and explosion, the origins of static electricity and the fundamentals of gas discharges. Chapter 4 summarizes the principles of safety and acts as a bridge between the fundamentals and the case histories which follow in Chapters 5–10. This latter section is the core of the book and consists of over 55 different descriptions of real cases. The authors hold the reader's interest in what could otherwise be a dry listing of facts by presenting the material much in the manner of a mystery story. They describe the individual cases by presenting different clues, some of which are seemingly insignificant, systematically evaluating them and then by leading the reader through appropriate deductions finally conclude the cause of the problem. In this way the reputed mysterious nature of electrostatics is made clear. The examples cover a broad range of industries and involve hazards associated with solids, liquids and powders. They range from the common to the esoteric. In many the truth seems stranger than fiction and yet all cases cited are based upon fact and are described exactly as they happened.

This book appears at a particularly appropriate time. Modern materials used in industry tend to have properties which are leading to more rather than fewer problems due to static electricity. By sharing their experience through presenting such a broad cross-section of examples the authors provide an invaluable aid for practitioners in industry to prevent many future accidents.

G. S. P. Castle

# Acknowledgements

The authors wish to thank Sylvia, who is Günter's wife, for her invaluable contribution to the writing of this book. Her ideas on its form of presentation and the sensitive manner in which she has helped in expressing some of the more difficult concepts and interrelations of electrostatics in readily understandable terms is very much appreciated.

It is also with pleasure that the authors acknowledge the debt they owe to their mentors, Dr G. Heyl, formerly of Bayer AG, Leverkuson, Germany and to the late Dr P. S. H. Henry, formerly of the Shirley Institute (now the British Textile Technology Group), Manchester, UK, for introducing them to the subject of static electricity and for their generous encouragement.

# Chapter 1
# Danger of fire and explosion

## 1.1 Basic considerations

Fire, explosion and detonation are forms of combustion which generally differ in their development. Fire is characterized mainly by a stationary burning flame in an open atmosphere and leads to an explosion when an ignition occurs in a combustible atmosphere within an enclosed space (e.g. a tank). Starting from the ignition source the flame-front runs through the entire space with a speed of about 10 m/s. The heating effect of the flame causes a pressure of about 10 bar (1.0 MPa) to build up, which diminishes during subsequent cooling. If the ignition starts in a rigidly confined space (e.g. a pipe) the heat of reaction causes a pre-compression which leads to a fast explosion (detonation) with a flame-propagating speed of up to 1 km/s. In this case the pressure may rise to 100 bar (10.0 MPa) resulting in a shot-like noise and, quite often, devastating damage.

## 1.2 Conditions for ignition

The combustion of fuel in air occurs between the tiniest particles of the fuel, i.e. the molecules, and oxygen. With liquid fuels the molecular fuel/oxygen mixture is easily produced by the vaporization of the liquid fuel. However, in the case of solid fuels it is necessary to break their chemical bonds so that their hydrocarbon molecules are set free to react with oxygen. Basically, a fire or an explosion will occur when fuel, oxygen and an ignition source exist together in what is known as a 'danger triangle' (Fig. 1.1). On the other hand, if one of these components is absent then the danger is removed. Although the latter is true it should, at the same time, be realized that the presence of the three components does not necessarily lead to combustion. This can be demonstrated by a simple experiment. When a lighted match is dropped into kerosene it becomes extinguished without causing an ignition. However, when the kerosene is heated to 45°C, the lighted match causes an ignition and the liquid continues to burn at its surface.

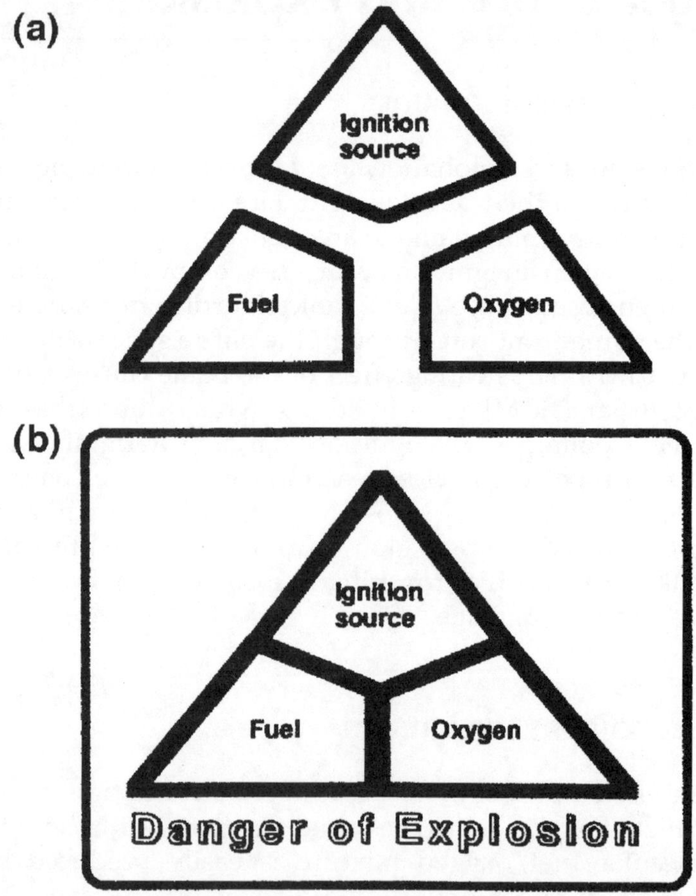


**Figure 1.1** (a) Components required for an explosion, (b) Danger triangle

It is, therefore, important that the interrelating factors governing the onset of combustion be properly understood.

## 1.2.1 Fuel

While it is necessary to distinguish between gaseous, liquid and solid fuels, a common feature between them is that combustion is sustained only within a certain explosion range, which is determined by the lower and upper explosion limits. For gases and vapours the limits are expressed in terms of the volume concentration (per cent) of the fuel in the fuel/air mixture. For combustible dusts the corresponding units are the mass of dust per unit volume of dust/air mixture (kg/m$^3$).

### 1.2.1.1 *Liquid and gaseous fuels*

The explanation for the behaviour of the kerosene in the above experiment has to do with the vapour pressure of the liquid. Depending on the temperature of the liquid a certain vapour pressure, and hence vapour concentration, is developed above the surface of the liquid. In Fig. 1.2 is shown the vapour pressure/temperature curve for ethanol and also the relation between the vapour concentration at the surface of the liquid and temperature. By using the curve, temperatures can be assigned to the lower and upper explosion limits of the liquid. The temperature (°C) related to the lower explosion limit is called the flash point and is a simple and reliable way of defining the danger of liquids from the point of view of their ease of ignition. Liquids at a temperature lower than their flash point cannot be ignited. The flash points for all flammable liquids are defined and listed in, for example, safety data sheets. Kerosene has a flash point of 45°C, indicating that it will not burn at room temperature.

Gaseous fuels, in general, show the same properties as liquid fuels but in a lower temperature range.

### 1.2.1.2 *Solid fuels*

In contrast to gases and vapours, mixtures of solid fuels with air are inhomogeneous because of the effect of gravity on the particles, e.g. with dusts in air the particle distribution is not constant with

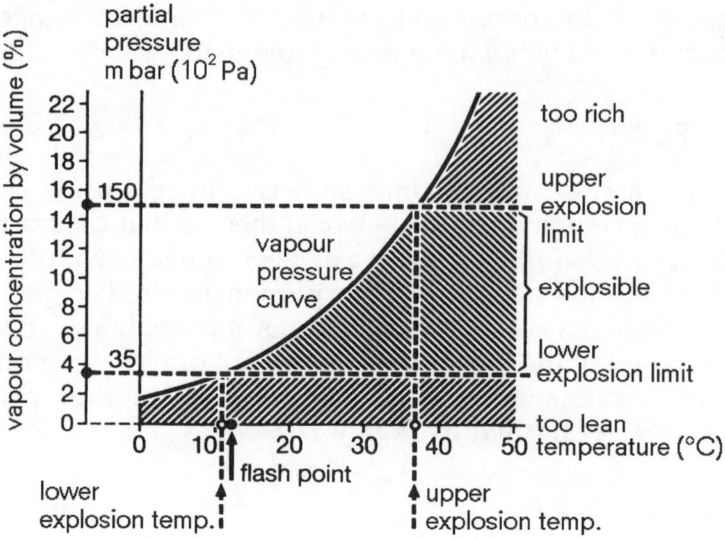


**Figure 1.2** Vapour concentration/pressure of ethanol as a function of temperature

reference to time and space. In terms of safety the explosion limits for dust/air mixtures are not as critical as those for vapour/air and gaseous/air mixtures. For instance a few millimetres of combustible dust settled on the floor may present an explosion hazard when swirled up by a draught of air.

## 1.2.2 Oxygen

In the following all statements relating to aspects of safety refer to atmospheric conditions unless otherwise stated. Among other things, this means that the oxygen content of the air is 21 per cent by volume. It is well known that oxygen-enriched atmospheres, found only in man-made systems (e.g. autogeneous welding), can be very dangerous. However, with regard to plant safety, it is also important to know the combustion properties of fuels in oxygen-diminished atmospheres. For all fuels a minimum oxygen concentration may be determined below which no combustion can occur. This limit is about 10 per cent by volume for most fuels. It follows from this that in industrial practice approximately 50 per cent of the air has to be substituted by an inert gas (e.g. carbon dioxide, nitrogen, etc.) in order to eliminate the danger of an explosion. For reliable safety, where there is doubt about the oxygen concentration, the latter should be determined *in situ.*

### 1.2.3 Ignition sources

Ignition sources are, according to scientific knowledge and experience, the means of releasing energy which is capable of igniting certain combustible materials when mixed with air. By the early 1960s, the evaluation of innumerable fire and explosion events had already shown that there were only 13 different ignition sources to be considered. Since then various experts have experimented with ignition sources but have found it impossible either to reduce the number by combining ignition sources of the same nature, or to find new ones. Today, 30 years later, the efforts of many experts throughout the world confirm that there are, indeed, only 13 ignition sources to deal with. They are listed below with short practical examples.

| **Ignition source** | **Example** |
|---|---|
| 1 Hot surface | heating pipe; casing of an electrical apparatus |
| 2 Flames and hot gases | autogeneous welding; exhaust gases |
| 3 Mechanical sparks | abrasive cutting; flint gas lighter |
| 4 Electrical equipment | electrical sparks at make and break |
| 5 Cathodic protection against corrosion, transient current | sneak current; short circuit to earth |
| 6 Static electricity | spark discharge; brush discharge |
| 7 Lightning stroke | |
| 8 Electromagnetic waves (high frequency range) | induction heating; radiotelephone |
| 9 Electromagnetic waves (optical range) | photoflash; laser |
| 10 Ionizing radiation | X-rays; UV-rays |
| 11 Ultrasonics | ultrasonic cleaning; ultrasonic testing |
| 12 Adiabatic compression | heat of compression; drift wave |
| 13 Chemical reaction | exothermic process |

The above list is taken from Explosionsschutz-Richtlinien EX-RL (1), which, it should be noted, does not rank the ignition sources according to their frequency of occurrence.

With regard to the ignitability of the ignition sources, there are some which are capable of igniting all kinds of combustible materials (e.g. flames, lightning stroke). However, it is different with hot surfaces, mechanical sparks and static electricity. These can only ignite certain combustible materials, depending on particular parameters, such as the ignition temperature and the minimum ignition energy (MIE) of the material.

### 1.2.4 Summary

The presence of fuel, oxygen and an ignition source (danger triangle) alone is not sufficient to cause a fire or an explosion; they have to interact in a prescribed manner, as is shown in Fig. 1.3, i.e.

- the combustible mixture has to be within the explosion limits
- the oxygen concentration must be higher than about 10 per cent
- the ignition source must release at least the appropriate MIE.

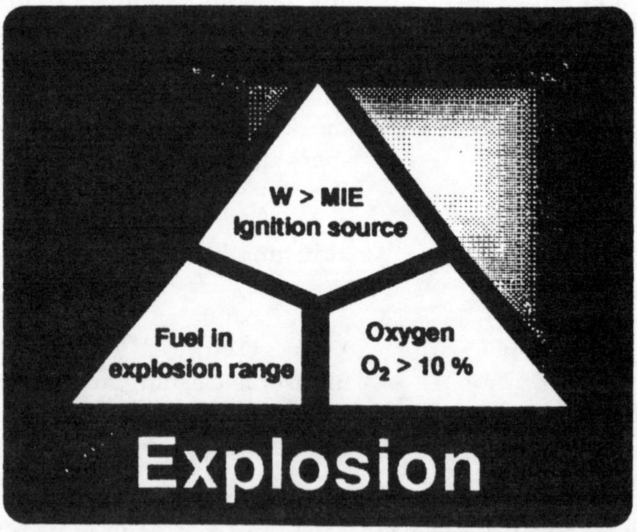


**Figure 1.3** Conditions for an explosion

## 1.3 Minimum ignition energy

The minimum ignition energy (MIE) of an optimum mixture of a combustible material with air (or oxygen) is defined as the least amount of energy needed to cause the ignition of the mixture when measured by a standard method. It is a means of classifying hazardous situations where fires and explosions might be initiated. The energy can be supplied in a number of ways (see 1.2.3) but is directly quantifiable only when delivered in the form of an electrostatic capacitative spark discharge.

The definition of MIE given above takes no account of the spatial and temporal distribution of the energy. The conversion of a given amount of electrostatic spark energy into heat could occur in a large volume and/or over a long period of time. Doubtless such conditions would be far less favourable for the promotion of an ignition than if the same energy was released into a tiny volume and/or over a short period of time.

A further complication is that not all of the energy released by the capacitor in a spark is converted into thermal energy. Some of the energy is lost as heat in the wiring of the discharge circuit and to the electrodes across which the spark passes; some is lost in the form of light and electromagnetic radiation and some by the pressure exerted by the spark. Also, there is always a small residual charge left on the capacitor after the discharge. Thus, the determination of MIE is, intrinsically, prone to error and so the measurement of precise values is not possible.

For an ignition to occur the concentration of the combustible material (gas, vapour, or powder) in the mixture must lie between an upper and a lower flammability limit. For concentrations above the upper limit there is insufficient oxygen to support and propagate combustion, while for those below the lower limit there is insufficient fuel for combustion. A plot of ignition energy against the concentration of the fuel in a fuel/air mixture is, typically, a U-shaped curve on which the lowest point denotes the MIE of the mixture. For gases (and vapours) the concentration is measured in terms of the volume of gas in the gas/air mixture, in per cent. For powders it is the mass of the powder per unit volume of the powder/air mixture. In terms of the ease of ignition and efficiency of burning of a gas it is the stoichiometric mixture which requires the minimum ignition energy, i.e. one in which there is the correct chemical balance between the gas and the oxygen in the air. Mixtures on either side of the stoichiometric one are either too lean or too rich in gas and, therefore, require more energy to be ignited.

Knowing the chemical equation for the reaction between a pure gas and oxygen, the volume concentration of the gas in the gas/air mixture can be calculated for a stoichiometric mixture. However, in practice it is often the case that the gas is not pure, and that even with a pure gas the optimum concentration is shifted away from the stoichiometric value during combustion. This arises because of the different rates of diffusion of the gas and oxygen, depending on their relative molecular weights, into the zone of the mixture which is about to be burned. In cases where the rate of diffusion of the gas is higher than that of the oxygen, because of its lower molecular weight, the mixture in the pre-burned zone is rich in gas and vice versa. To compensate for a pre-burned zone which is rich in gas the original mixture has to be on the lean side of the stoichiometric value, while for one which is lean in gas the mixture has to be on the rich side of the stoichiometric value.

By plotting ignition energy against the concentration of fuel in the fuel/air mixture as a fraction of the stoichiometric mixture for benzene and methane (Fig. 1.4) it can be seen that the MIEs of the fuels do not coincide with their stoichiometric mixtures. With methane the MIE is for a mixture which is on the lean side of the stoichiometric value. This is expected because the molecular weight of methane is less than that of oxygen. On the other hand, the MIE for benzene is for a mixture which is on the rich side of the stoichiometric value because the molecular weight of benzene is greater than that of oxygen.

## 1.3.1 Determination of minimum ignition energy

### 1.3.1.1 *Gases and vapours*

In measuring the MIE of a gas or a vapour there are a number of factors which need to be considered relating to the experimental procedure. From the discussion in 1.3 it follows that, usually, it will be necessary to adjust the concentration of the fuel in a fuel/

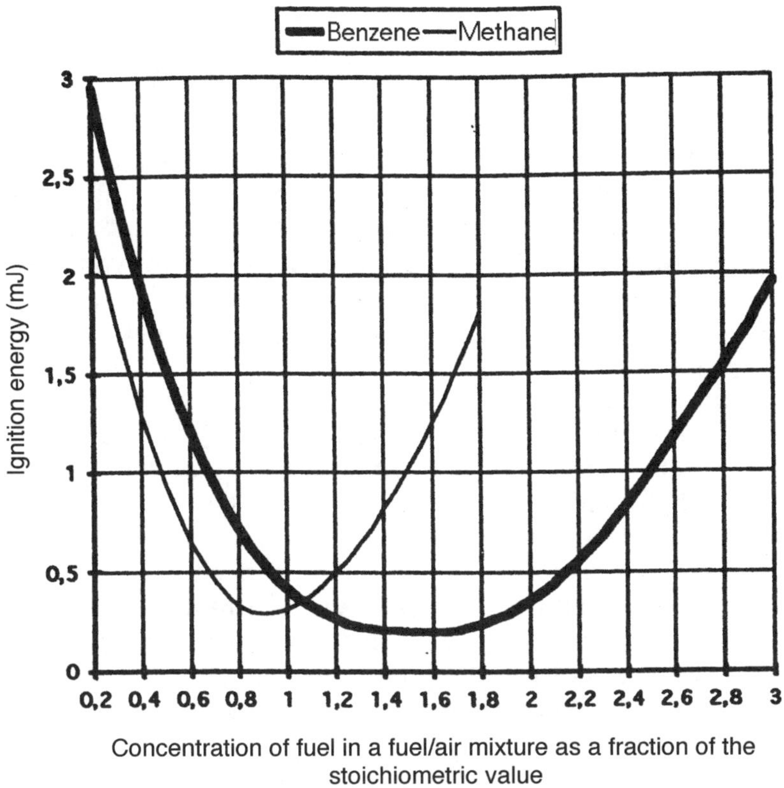


**Figure 1.4** Dependence of ignition energy on fuel concentration

air mixture until the mixture can be ignited by using the minimum energy.

The ignition source is an electrostatic capacitative spark discharge across two electrodes located within the mixture under test. It is important that the size and spacing of the electrodes are optimal if the conditions for minimum ignition energy are to be achieved. The diameter of the electrodes should be large enough to avoid corona discharges and their spacing just wide enough to avoid quenching of the expanding flame kernel due to thermal losses at the electrodes by conduction. For electrode spacings below the critical quenching distance the energy for an ignition rises steeply as more energy is needed to compensate for the losses to the electrodes. Electrodes which are spaced more widely than the quenching distance require a greater potential across them to cause a spark discharge. The size of the capacitor supplying the spark energy has also to be carefully selected. It must be just large enough to provide the required MIE when at a potential equal to the breakdown potential of the gap between the electrodes. For example, most hydrocarbon gases mixed with air have MIEs in the region of 0.2–0.3 mJ and their critical quenching distances are about 2 mm. Taking the lower energy value as the MIE of the gas concerned, and assuming a breakdown potential of, say, 7 kV, the corresponding capacitance, calculated by using the energy equation (see 2.5.12), is about 8 pF.

A method for producing capacitative spark discharges of known energy is described in BS 5958: Part 1: 1991 (2). A capacitor and an electrostatic voltmeter are connected in parallel across two electrodes mounted inside an ignition chamber. One of the electrodes is grounded and the voltmeter is isolated from the capacitor by means of a decoupling resistor. The capacitor is charged up slowly by a high voltage DC supply via a current-limiting resistor until a spark occurs across the electrodes. The cycle continues, giving a series of sparks at regular intervals. The potential on the voltmeter at which an ignition occurs is noted. The energy equation is then used to determine the energy of the spark between the electrodes.

#### 1.3.1.2 *Dusts*

Equipment for determining the MIEs of dusts is described in BS 5958: Part 1: 1991 (2). However, it is well known that accurate measurements are extremely hard to achieve, partly because of the difficulty of ensuring a uniform distribution of the dust in the ignition chamber at the time of the spark discharge, and partly because of the contamination of the electrodes by the dust itself.

More work is required in this field if a satisfactory method for measuring the MIEs of dusts is to be established.

### 1.3.2 Dependence of MIE on oxygen concentration

It has been pointed out in 1.2.2 that when the oxygen concentration within most fuel mixtures falls below about 10 per cent by volume no combustion can occur. This limitation of the oxygen concentration is a very important safety measure which is applied mainly in the chemical industry. As 10 per cent is only an indicative value it is necessary, when reliable inerting is required, to determine the exact value of the oxygen-limiting concentration (OLC). The latter is measured for an optimum fuel concentration using a very strong ignition source which supplies spark discharges of energies in the range 2–10 kJ, depending on the volume of the test apparatus.

When comparing the definitions of OLC and MIE it follows that both characteristic values indicate limits with respect to ignition and non-ignition. When determining the MIE of a fuel using oxygen from the air, the oxygen concentration is fixed and the energy of the spark from the igniting source has to be varied. On the other hand, when determining the OLC the energy is fixed (at a high level) and the oxygen concentration has to be varied.

With this in mind Glor and Schwenzfeuer (3) began investigations into the effect of oxygen concentration on MIE.

From experiments with combustible dusts it was found that small reductions in the oxygen concentration led to large increases in the MIE. For example, coal dust has a MIE of 0.1 J under atmospheric conditions (oxygen concentration of 21 per cent), but when the oxygen concentration was reduced to about 18 per cent the MIE increased to 1.0 J.

This fact has a large influence on electrostatic ignitions as will be shown later on. In anticipation, it can be said here that electrostatic discharges are generally very weak (excepting propagating brush discharges) and do not reach energy values of 1.0 J. This means that a small reduction in the oxygen concentration of the air can prevent electrostatic ignitions.

### 1.3.3 Overview

This book deals with a large variety of fires and explosions which have been caused mainly by static electricity. A number of cases arising from causes other than static are also presented (Chapter 9). Nevertheless, when considering the total number of accidents it becomes evident that static electricity makes only a minor contribution to the cause of ignitions. This is despite the fact that static charges are found everywhere and are, certainly, part and parcel of everyday human activity. In this context it has to be explained why, for example, in an industrial plant where the conditions may have remained unchanged for a long period of time there is quite suddenly an electrostatic ignition!

By referring to the explosion triangle (Fig. 1.3) it can be seen that an explosion will occur when a fuel/air mixture, which is in the explosion range, is present together with an ignition source of energy equal to or greater than the MIE of the mixture. As plant managers generally endeavour to avoid explosible atmospheres and possible ignition sources, explosions are fortunately rare incidents. Most electrostatic ignition sources are quite weak and hardly reach the MIE levels required for, say, saturated hydrocarbon gases and vapours and other fuels of similar MIEs. They are, therefore, unable to ignite flammable mixtures which are within the explosion range including those of the highest sensitivity to ignition. As is shown in Fig. 1.4 the MIEs of benzene and methane are approximately 10 per cent of their corresponding upper and lower limit ignition energies.

The main contents of this book are the case histories. It is noticeable that toluene, in particular, is involved in a disproportionately large number of cases related to brush discharges (Chapter 5) and spark discharges (Chapter 8; Liquids). This is not just by chance. The flash point of toluene is 6°C which means that explosible mixtures with air are formed at this temperature and above. However, at room temperature (about 20°C) the mixture is close to the most easily ignitable one. Thus, processes involving toluene at room temperature give rise to a sensitive flammable atmosphere which may be readily ignited by electrostatic gas discharges. This is, in fact, indicative of why ignitions caused by static electricity are so rare. The statistically scarce event of an electrostatic ignition source coinciding with the presence of an explosible atmosphere is further diminished by the fact that only certain fuel concentrations are ignitable. In a way this is a tantalizing and exasperating aspect of electrostatic hazards because, although accidents occur very rarely, it would be inviting disaster not to take them into account when planning for safety.

## 1.4 Literature

(1) Explosionsschutz-Richtlinien EX-RL (1996). Hauptverband der gewerblichen Berufsgenossenschaften, Lindenstr. 78–80, D-53757, Sankt Augustin.
(2) BS 5958 (1991). Code of practice for control of undesirable static electricity. Part 1. General considerations.
(3) Glor, M. and Schwenzfeuer, K. (1996). Sichere Handhabung brennbarer Staube. VDI-Berichte 1272.

# Chapter 2
# Origin of static electricity

Man-made current electricity has been in existence now for about a hundred years and its use as a source of power in industry, commerce and in the home is widespread throughout the world. The origins of current electricity are to be found in static electricity – a phenomenon, which although known to man for thousands of years, is today still considered to be something of a mystery. A contemporary Englishman has succinctly described it thus: 'Static electricity is an unpredictable phenomenon, here today and gone the next.'

It is probably because of its unpredictability that static electricity is often, incorrectly, blamed as the cause of industrial accidents involving fire and explosion when no other plausible explanation can be found.

Electricity is a manifestation associated with the fundamental electrical charge to be found within the atom, namely, the electron. A surplus of electrons, which are negative charges, results in a region of negative polarity. A deficiency of electrons signifies positive charges, and a region of positive polarity.

In deciding how to approach the subject of charge formation we consider the analogy of someone going on holiday. The person would need to take with him a suitable map, e.g. in the case of a hiker, a map on which the contours and footpaths were shown in detail, and for a motorist, one on which the roads were given prominence. In a way a map can be regarded as a model of the area to be explored. An analogy in electrostatics could be the exploration of the formation of static electricity. Some explorers might be interested in the fundamental quantum mechanics of charge transfer, as is described in the electron energy band model, while others would choose a phenomenological approach.

For the practical cases which are dealt with in this book, the latter is considered to be the most useful and is described in the following section.

## 2.1 Double layer charge

Over the centuries many views have been expressed on how electrostatic charging comes about, and it was not until the advent of semiconductors that a theory was established which has gained general acceptance. The theory is based on the fact that electrons are emitted from solid surfaces when their temperature is high enough, e.g. thermionic emission from a metal surface in a vacuum, as in a cathode ray tube. The energy required to remove an electron from the surface of a material to infinity is called the work function. It is different for various materials and is diminished by the application of heat. The work functions of insulating materials (e.g. plastics) are high but for materials with plenty of conducting electrons (e.g. metals) they are low.

When two different materials at the same temperature are brought into firm contact with each other, i.e. with a distance between them in places of the order of a few nanometres, a transfer of electrons takes place across the interface, the number of which is dependent on the difference in the work functions of the two materials. A double layer of charge is produced at the interface as is described by Helmholtz (1). In terms of this simple representation, electrons from the material (donor) with the lower work function migrate to the material (acceptor) with the higher work function thereby producing a layer of negative charge on one surface and positive charge on the other. An equilibrium state is reached when the potential difference corresponding to the two work functions equals the potential difference between the two layers of charge of opposite polarity. In the example given in Fig. 2.1 the work function of material 2 is lower than that of material 1. Consequently, electrons move from 2 across the

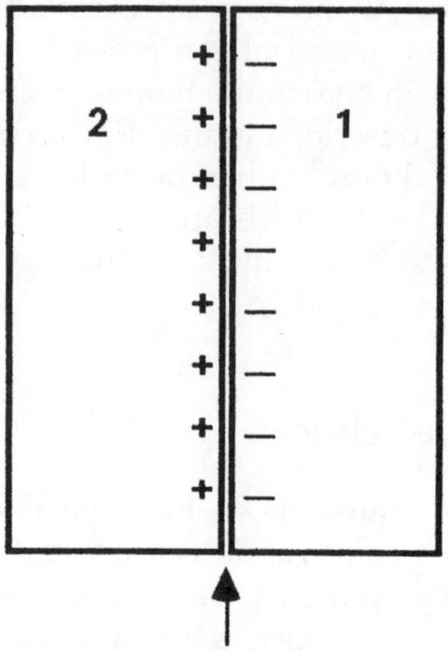


**Figure 2.1** Charge separation between contacting materials. Donor material 2 has a lower work function than acceptor material 1

interface to 1, causing 1 to become charged negatively and leaving 2 in a positively charged state. This effect is known as contact electrification (contact charging). The potential difference caused by this polarization of charge at the interface is usually of the order of millivolts and, because of the extremely small gap between the surfaces, the capacitance of the system is very high. Contact charging itself depends only on surface contact between materials irrespective of whether or not the surfaces are stationary or moving relative to one another.

The term frictional electrification (frictional charging) has been in use for many years with the implication that it is a different process from contact electrification. However, it is now known that friction in itself has no bearing on the electrification process. Basically, all that it does is cause an increase in the area of contact between the surfaces. Despite this, the term triboelectrification (tribo-charging) is still used to describe occurrences which are actually contact charging.

## 2.1.1 Charge separation

Following their contact electrification the surfaces are moved apart (Fig. 2.2) so that the distance between them increases by several orders of magnitude. This is accompanied by an enormous reduction in the capacitance of the system and corresponding increases in the potential difference between the surfaces and in the energy of the charge in the system. However, the question remains as to what is the nature of the charges taking part in the charging process. From much experimental work on contact and tribo-electrification three different types of charge carrier are identified, namely, electrons, atmospheric and surface ions, and charged particles of the materials making contact. Experiments done on contact electrification with clean materials in vacuum indicate that electrons are the main contributors to the charging process. However, charging between similar materials in a practical situation, e.g. when a plastic film is unwound from a roll, seems to be dependent largely on the transfer of ions across the surfaces.

In separating two surfaces the coulomb forces of attraction between the opposite layers of charge at the interface have to be overcome by expending mechanical energy on the system. This energy is instantly converted into electrical energy and it is then

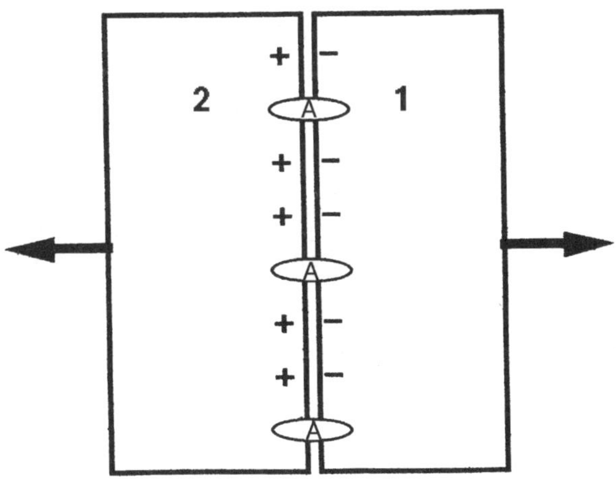


**Figure 2.2** Materials 1 and 2 moving apart. Charge is neutralized at the last points of contact A

that the electric fields from the exposed charges can become effective in causing gas discharges and electrostatic induction.

Any consideration of contact charging must always be based on the charges transferred between the different materials in close contact. The interaction of this charge with the drastically reduced capacitance of the system, as the surfaces move apart, leads to a high potential difference between the materials. In practice the total charge transferred during contact is reduced as the surfaces are separated primarily as a result of two factors, namely, surface resistivity and gas discharge.

*Surface resistivity* During the separation of two surfaces the charge on each of them has a tendency to flow across the surfaces to the last points of contact, as indicated by A in Fig. 2.2, where the charges are neutralized. The speed with which this partial discharging of the surfaces by conduction occurs depends on the resistance of the path taken by the charges. The greater the resistance the slower is the discharging and vice versa. It may be deduced from this that with high speeds of separation of the surfaces (short times for discharge) and/or high surface resistances, the neutralization of charge by conduction will be restricted and a larger amount of the original charge will remain on the surfaces. On the other hand, with low speeds of separation (long times for discharge) and/or low surface resistances, the charges on the surfaces will be readily neutralized thereby leaving little charge behind. It also follows that the charging of the surfaces can only occur if at least one of the materials being separated is of a high resistance.

*Gas discharges* As the surfaces in contact are separated, the electrical field strength in the gap may reach the breakdown value for air (3 MV/m) resulting in gas discharges. The ions which are produced are attracted to opposite charges on the surfaces which they neutralize. This effect occurs, for example, when removing a woollen sweater from the body or when adhesive tape is pulled off a roll, and can be accompanied by audible 'cracks' and, in the dark, a visible glow. Because of the reduction in charge described above it is not possible to predict the amount of charge remaining (Fig. 2.3) after the surfaces are separated.

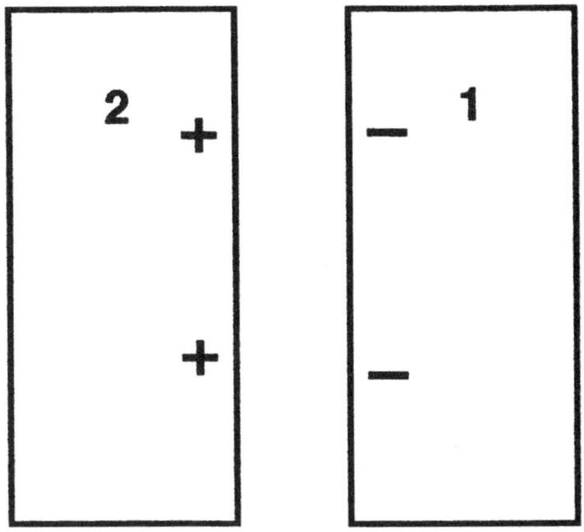


**Figure 2.3** Residual charge on materials 1 and 2 after their separation

## 2.2 Charging of liquids and gases

As with the static electrification of solids, it is important to understand also the charging behaviour of liquids and gases, particularly in the context of their manufacture and use in industry.

### 2.2.1 Charging of flowing liquids

Double layers of charge are formed, as with solids, at liquid/liquid and liquid/solid interfaces. However, in contrast with the charging of solids, that for liquids also requires the additional presence of ions. For the case, say, of a liquid being transferred to a vessel via a metal pipe, if positive ions from the liquid are deposited on the surface of the pipe, then negative charges of equal magnitude will form a diffuse layer in the liquid. These negative ions are passed with the liquid into the vessel. This is equivalent to the vessel being charged by an electric current. The amount of charge swept along in the liquid depends on the volume resistivity of the liquid and the speed at which the liquid is separated from the pipe, in a similar way to the case with solids. It should be noted that the conducting properties of liquids are normally expressed in terms of their electrical conductivity which is the inverse of their resistivity. In general, the means of preventing or reducing the charging of liquids are the same as those recommended for solids.

The formation of interfacial double layers of charge may also occur at the inner interfaces of dispersions and emulsions. With these systems large interfacial areas between the components are possible where high amounts of static electricity can be anticipated, e.g. with an emulsion of a few per cent of water and an insulating hydrocarbon liquid, as has been described by Klinkenberg and van der Minne (2).

### 2.2.2 Charging of flowing gases

As has already been discussed, contact electrification can occur at solid/solid, solid/liquid, and liquid/liquid interfaces. In the case of gases there is no static electrification at their boundaries. This means that gases flowing in pipes will not themselves give rise to static charges. However, any solid or liquid particles (dusts and aerosols) entrained in the gas stream may readily produce charges. In some cases, e.g. with compressed air, aerosols may be present in the form of condensed water or oil mist which have come from the compressor.

## 2.3 Reducing the tendency of charging

The following methods may be used as means of minimizing the formation of static charges.

### 2.3.1 Decrease the interfacial area of contact

From the discussions above it is clear that electrostatic charging is, essentially, a surface phenomenon. It therefore seems reasonable to expect that the formation of charge may be reduced by decreasing the area of contact between the materials to be separated. The effect of moderately roughening surfaces by mat finishing them is known to significantly reduce static charging, e.g. shafts used in rewinding machines.

## 2.3.2 Use of the triboelectric series

As has already been noted, the separation of electrostatic charges depends, primarily, on the difference in the work functions of the materials involved. The greater the difference, the larger is the charge separated and vice versa. On this basis it should be expected that to reduce static charging materials should be selected which differ as little as possible in their work functions. Indeed, the contact electrification between identical materials should be zero. Coehn (3) put forward a rule, stating that materials of high relative permittivity (dielectric constant) become positively charged when separated from materials of low permittivity. A corresponding triboelectric series is shown in Table 2.1. Various other series have been established experimentally for a wide range of materials, and the work of, for example, Montgomery (4) and Unger (5) is typical in that they indicate only small differences in the placing of given materials in the series. The series is defined

**Table 2.1** Triboelectric series (after Coehn) (abbreviations: see Table 2.2)

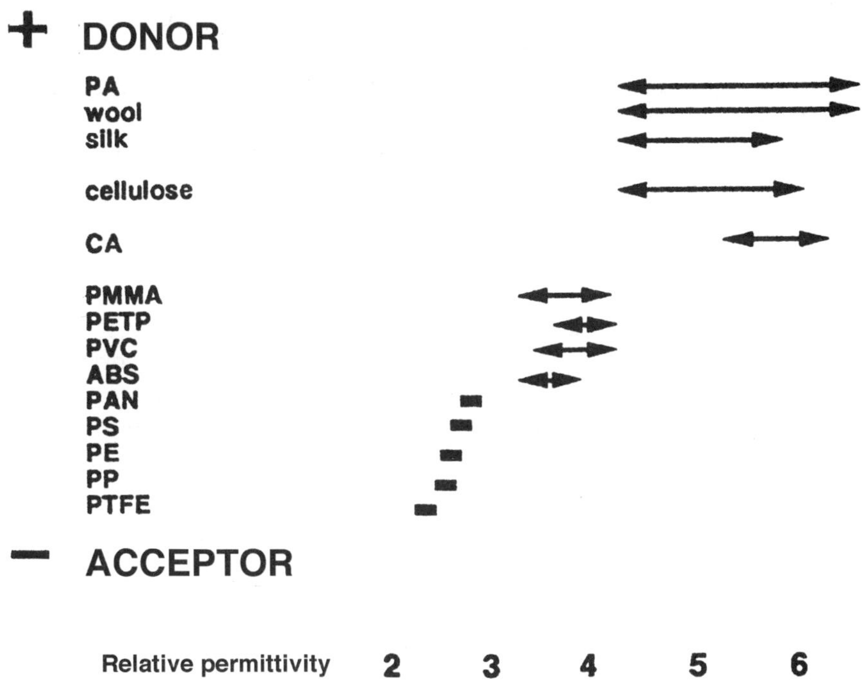

as a list of materials arranged in such a way that any one of them becomes positively charged after contact with any material below it in the series. However, the magnitude of the charges on pairs of materials after contact and separation does not always correlate with their distance apart in the series. Indeed, charge is usually produced when nominally identical materials make contact with each other and are then separated. It is likely that these departures in behaviour from that expected from theory are caused by the presence of impurities and adventitious ions on the surfaces of the materials.

Another possible contributing factor is that of asymmetrical rubbing. When materials are rubbed together the area of each surface rubbed is not the same. An extreme example of this is when the bow of a violin is drawn across a string. Although in this case the contacting materials are of the same type, namely, collagen, the gut string always becomes positively charged and the horse hair of the bow negatively charged. This phenomenon was explained by Henry (6) as follows. As the bow is drawn over the string a much larger area of the horse hair is rubbed than that of the gut. Thus the frictional energy and the heat produced is concentrated at the place on the gut over which the bow slides. This leads to a temperature gradient in which the gut is warmer than the bow. The electrons in the gut are, therefore, in a higher energy state than are those in the horse hair and so move down the temperature gradient, thereby causing the gut to become positively charged and the horse hair negatively charged.

### 2.3.3 Reduce the surface resistivity

Without doubt, the most reliable means of reducing the separation of static charges between two materials is to decrease their surface electrical resistivities to an appropriate level. However, as has been pointed out, the speed of separation of the surfaces is also a factor which influences the amount of charge recombination. Therefore, the question arises as to which threshold value of resistivity does static become a problem. This has been investigated for the case of normal speeds of separation which are in the region of 1 m/s. The charging behaviour of a large number of plastic materials has been measured according to DIN/VDE 0303 (7).

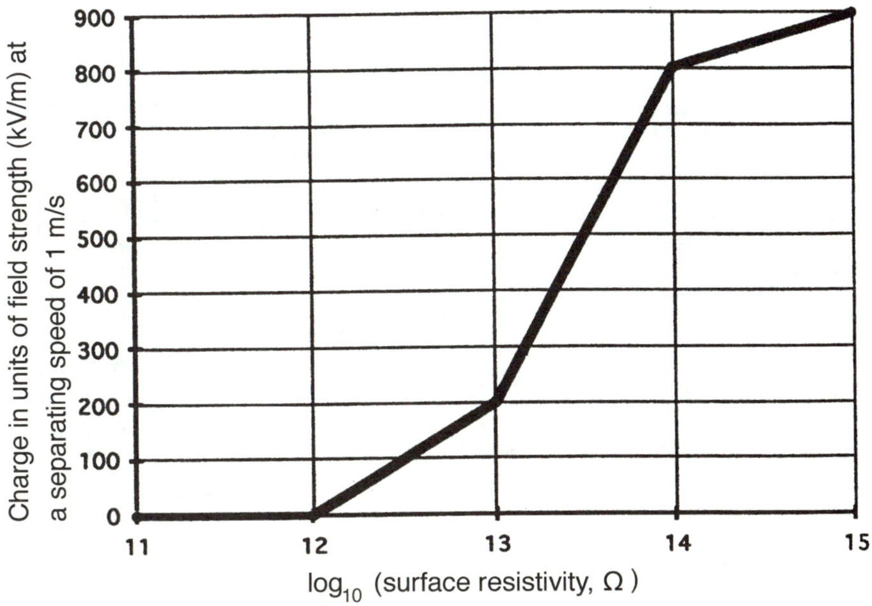


**Figure 2.4** Relationship between surface resistivity and chargeability of various plastics

The results given in Fig. 2.4 show that electrostatic charge, measured in terms of field strength, is only detected on surfaces of resistivity exceeding 1 TΩ at separating speeds of 1 m/s. Values of the surface resistivities of some typical plastic materials are given in Table 2.2.

An important factor which influences the surface resistivity of plastic and other materials is the relative humidity of the ambient atmosphere. By absorbing moisture from the air the surface resistivities of many plastics are reduced, as can be seen in the examples given in Fig. 2.5. It follows that tests for surface resistivity should be done under the prevailing conditions of use if they differ from the standard test conditions. However, it should not be assumed that an increase in relative humidity always causes a decrease in the tendency for plastics to become charged. For example, the polyolefins, which are a group of polymers with very low moisture absorbency, show only a very slight fall in their chargeability at higher humidities.

**Table 2.2** Surface resistivities of various plastics at 50 per cent r.h. and 23ºC

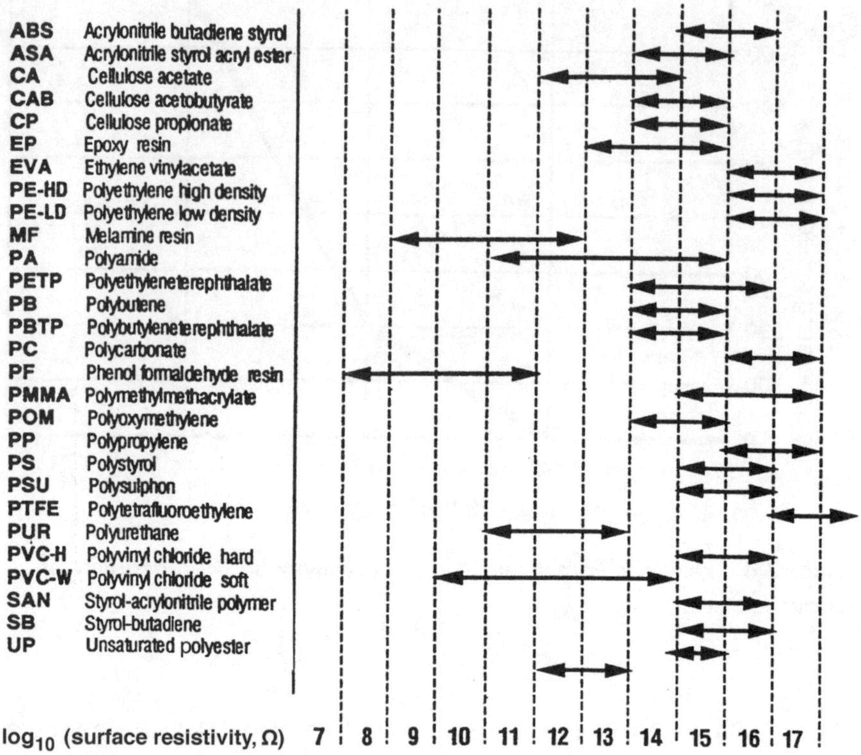


## 2.4 Electrostatic induction

An electric field has the ability to cause the movement of charges in nearby conducting bodies. The process, which is called electrostatic induction, causes a separation of charge within the conductor when it is placed in the field. The induction effect is illustrated in Figs 2.6–2.11. In Fig. 2.6 the sphere is part of a Van de Graaf generator and the rectangle facing it is a conductor. Both are mounted on insulating columns and are temporarily grounded. In Fig. 2.7 the generator is switched on for a few seconds to charge the sphere negatively. Under the influence of the electric field from the sphere electrons move to the far side of the conductor leaving the near side positively charged. On

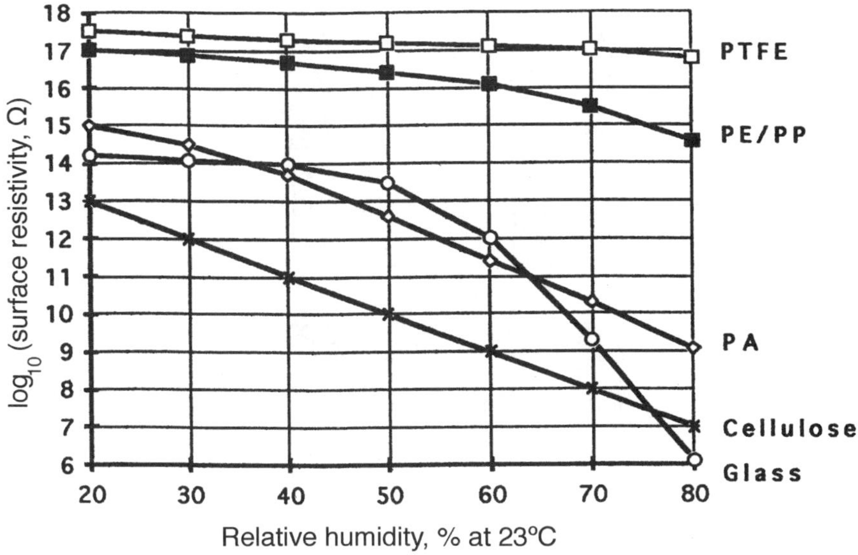


**Figure 2.5** Influence of air humidity on surface resistivity

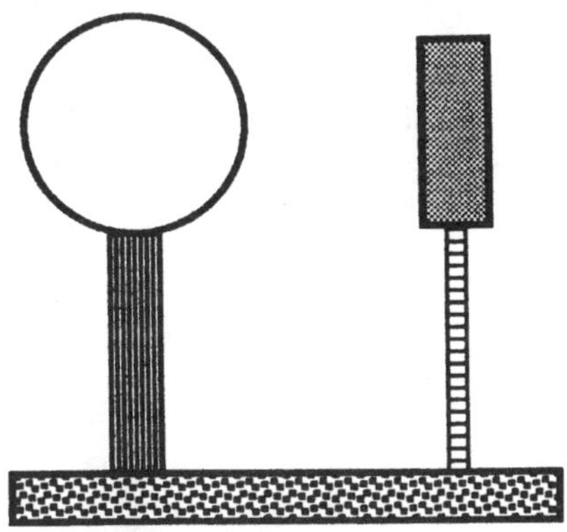

**Figure 2.6** Position of the neutral objects

grounding the conductor (Fig. 2.8) the negative charge is led away to ground while the positive charge is held captive by the negative charge on the sphere (Fig. 2.9). On removing the sphere (Fig. 2.10) the positive charge on the conductor is redistributed over its surface where it remains (Fig. 2.11).

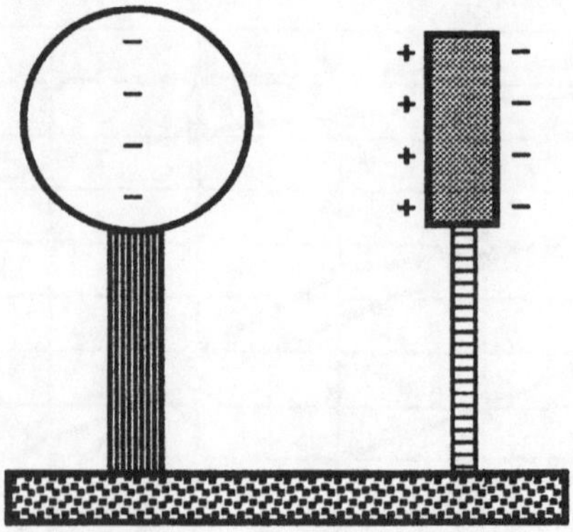

**Figure 2.7** Charge induced on the conductor by negative charge on the sphere

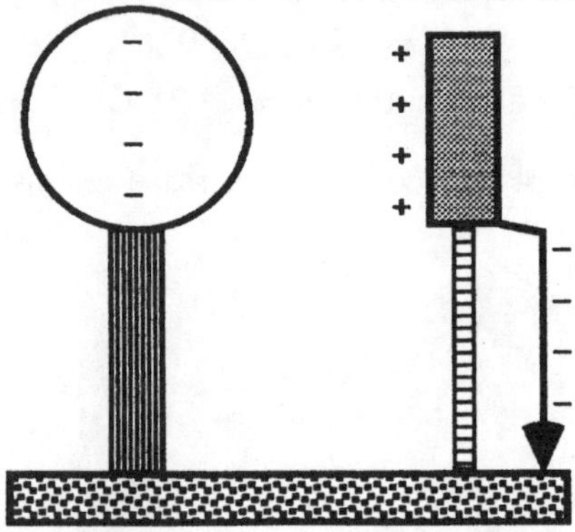

**Figure 2.8** Grounding the conductor causes induced negative charge to flow to ground

**Figure 2.9** Equivalent positive charge remains on the conductor

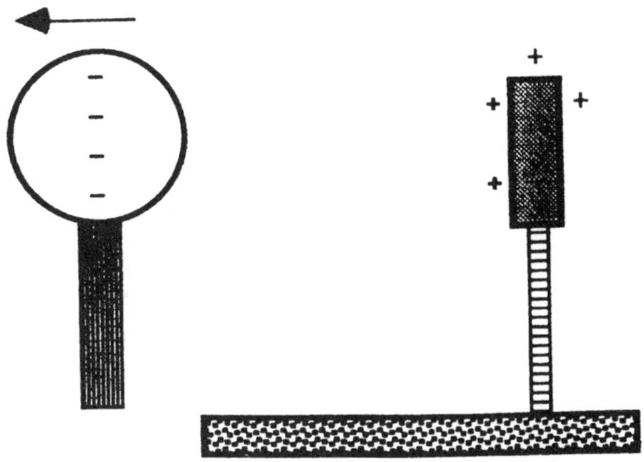

**Figure 2.10** Sphere removed; positive charge migrates over the surface of the conductor

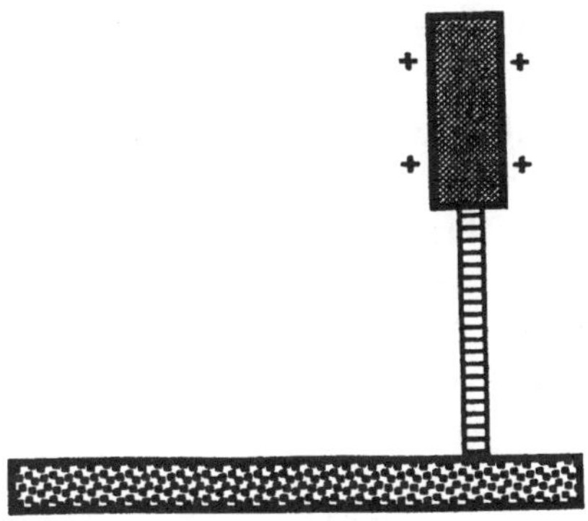

**Figure 2.11** Conductor remains positively charged

As a practical example, the effect of induced charge on the human body is now considered. When a person who is insulated from ground by, say, footwear approaches a highly charged surface, e.g. a roll of plastic film in the factory, a charge is induced on the person which results in an electrostatic shock on his touching a large or grounded conductor. When the person walks away from the roll an opposite charge to that initially induced on the body remains on the person which results in a second shock when the person again touches a grounded or large conductor.

## 2.5 Basic concepts and units

### 2.5.1 Charge

Symbol $Q$
Units: amperesecond (As) = coulomb (C)
Comment:

- Charges may have a positive or a negative sign.
- Charges give rise to electric fields.
- Charges exert forces on one another in accordance with Coulomb's law. The force is proportional to the inverse of the square of the distance between the charges, and to the product of the charges.
- Charge can neither be created nor destroyed in a self-contained system (Law of the conservation of energy).
- The smallest charge $e$ is that on the electron where $e = 1.6 \times 10^{-19}$ C.

Practical examples:

- A person of capacitance 200 pF and a potential of 10 kV carries a charge of 2 μC.
- The charge on pneumatically conveyed fine dusts can reach 1 mC/kg.
- A 100 W light bulb at 230 V passes a charge of about 0.43 C every second.

### 2.5.2 Surface charge density

Symbol $\sigma_e$
Units: C/m$^2$
Comment:

- There is a limit to the density of charge on a solid surface. Under normal atmospheric conditions the maximum charge density is 26 μC/m$^2$.

### 2.5.3 Volume charge density

Symbol $\rho_e$
Units: C/m$^3$

### 2.5.4 Potential

Symbol $\phi$
Units: volt (V) = joule/coulomb
Comment:

- The potential at a point in space is defined as the work required to bring a unit charge from infinity or ground up to the point.

### 2.5.5 Voltage

Symbol $U$
Units: volt = V
Comment:

- The potential difference between any two electrically conducting points in space is the voltage $U$. As long as the voltage specified is for one point in space, either ground or infinity are taken as being at zero potential.

Practical examples:

- The contact potentials between different materials are of the order of mV.
- When taking off a pullover the potential on a person insulated from ground may reach 10 kV.
- A Van de Graaf generator produces potentials of about 100 kV relative to ground.

### 2.5.6 Field strength

Symbol $E$
Units: V/m
Comment:

- An electric field is produced by electric charges.
- The direction of the field is out of a positive charge and into a negative charge.
- As electric fields can be superimposed according to the principle of superposition, they must be added vectorially. Any inhomogeneities in a material cause changes in the field strength.
- Field strengths of 3 MV/m, under normal atmospheric conditions, give rise to spontaneous gas discharges.

### 2.5.7 Current intensity (current)

Symbol $I$
Units: Ampere = A
Comment:

- Current intensity is the amount of charge flowing through a cross-section of material per unit time.
- The direction of the current is the same as that of the moving positive charges.

Practical examples:

- A Van de Graaf generator produces a current of about 20 μA.
- Charging currents produced by manual procedures in industry are usually less than 1 μA.
- Currents of up to 50 μA may be reached during the pneumatic conveying of materials.

### 2.5.8 Resistance

Symbol $R$
Units: volt/ampere = ohm (Ω)
Comment:

- For many materials, e.g. metals, the resistance is independent of the applied voltage, i.e. they obey Ohm's law. However, the resistances of poor conductors and of non-conductors can be strongly dependent on the applied voltage.
- When measuring resistance the appropriate standard test method should be used.

### 2.5.9 Volume resistivity

Symbol $\rho$
Units: Ω m
Comment:

- The volume resistivity of a material is calculated by multiplying the measured resistance by the ratio of the area of the electrodes to the distance between them. For liquids it is normal to use the term volume conductivity, which is the reciprocal of volume resistivity and has the units of $1/\Omega m$ or S/m, where S is in siemens.

### 2.5.10 Surface resistivity

Symbol $\sigma$
Units: $\Omega$
Comment:

- The surface resistivity of a material is calculated by multiplying the measured resistance by the ratio of the length of the electrodes to the distance between them. It is equivalent to the resistance between opposite edges of a square of the material and is independent of the size of the square.

### 2.5.11 Permittivity (free space)

Symbol $\varepsilon_0$
Numerical value and units: 8.85 pF/m = 8.85 pA s/V/m
Comment:

- The permittivity of free space $\varepsilon_0$ is related to the permittivity $\varepsilon$ of a dielectric and its relative permittivity $\varepsilon_r$ (dielectric constant) by the following equation, $\varepsilon_0 = \varepsilon/\varepsilon_r$.

### 2.5.12 Capacitance

Symbol $C$
Units: coulomb/volt = farad (F)
Comment:

- Any two electrical conductors which are isolated from each other form a capacitor. Its capacitance $C$ is given by the ratio of the charge $Q$ on either conductor and the potential difference $U$ between the conductors.
- Capacitance is a measure of the ability of a capacitor to store charge as expressed in the equation $Q = CU$. For a parallel plate capacitor the capacitance depends on the area $A$ of the plates and the distance $d$ between them. If there is a dielectric (polarizable) material between the plates of the capacitor, the capacitance

increases proportionally with the relative permittivity of the dielectric.

The capacitance $C$ of a parallel-plate capacitor is given by the equation, $C = \varepsilon A/d$.

- The rate of charging or of charge dissipation of a capacitor $C$ through a resistor $R$ is exponential with time. The product $RC = \tau$ is called the relaxation time or the time constant of the circuit, i.e. the time taken for the capacitor to charge up to $0.63U$ or for it to discharge down to $0.37U$.
- The electrical energy $W$ of the charge stored on a capacitor is determined by the equation $W = 0.5CU^2$ watt-second (joule)

Practical examples:

- Person wearing insulating shoes 100–200 pF
- Metal drum placed on an insulating support 200–400 pF
- Metal flange on plastic or glass tubes 10–30 pF
- Vehicle with insulating tyres 500–1000 pF

## 2.6 Static charges on the clothing and the body

Human beings, certainly during their waking hours, are almost invariably in an electrostatically charged condition. The charge is usually quite small and often goes unnoticed. In cases where a person is grounded, either directly or through conducting footwear and a conducting floor, the charge on the body is zero.

Larger charges are, most frequently, accumulated on the body when the latter is active. For example, when a person walks over a carpet or slides out of the seat of a motor vehicle. The removal of an outer layer of clothing will often cause the body to become charged. Charge on the body can arise by means of induction caused by electric fields from nearby charged surfaces (see 2.4) or by a person touching a charged object when the latter is conducting enough to be able to share its charge with the body. People working in dusty conditions become charged when charged dust settles on their clothing.

The problem of static charges on the clothing and the body is that they can give rise to static discharges from each of them. In this respect, the greater worry has been with spark discharges from the body which, under optimum conditions, can be incendive in the presence of combustible atmospheres. They may also cause shocks to the body but these are not usually a cause for concern. In recent years the effect of static discharges from the body on the electrical integrity of sensitive electronic components has been the subject of much work and discussion.

Gaseous discharges from clothing on the body have to be taken into account when they are in the presence of very sensitive combustible atmospheres such as hydrogen or acetylene mixed with air and in clean rooms where electronic components are handled.

As the ignition of very sensitive flammable materials and the damage caused to electronic components by gaseous discharges are outside the scope of this book, it is sufficient to note that there is, at present, no published information available on the observed effects of static discharges from clothing on the body relating either to the ignition of very sensitive gases and vapours or to component damage.

In terms of case histories, there are to our knowledge hardly any properly documented and authenticated examples on record of fires and explosions caused by static discharges from the body, let alone the clothing on the body. In the case histories to follow only one example of an incendiary discharge from the body is cited (see 8.26). However, this dearth of recorded accidents should not be taken as a signal for any relaxation in the implementation of the appropriate safety precautions when necessary.

The incidence of shocks to people caused by charge on the body are, these days, a fairly common occurrence.

Because of these situations it is desirable that the mechanisms involved in the electrostatic charging and discharging of people and their clothing be understood, if only to shed some light on why accidents arising from these sources are so rare and shocks so relatively common.

### 2.6.1 Static discharges from the body

#### 2.6.1.1 *Incendiary discharges*

Probably the most frequent means whereby the body becomes charged is by charge separation (see 2.1.1) at the interface between, say, the clothing and a seat cover or when the former brushes against other surfaces and other people. The field from the charge on the clothing instantly induces (see 2.4) a similar charge onto the body which is retained when the body is isolated from ground. Induced charges on the body are also caused by separated charges on the shoe sole as a person walks about and by the charge left on a sub-layer of clothing when an outer garment is removed from the body. The effect of these induced charges is to cause a change in the potential of the body, relative to ground, either positively or negatively depending on the polarity of the charge.

With the body in this condition an electrostatic spark can be produced, say, when a finger approaches a grounded or large conductor. The difference in potential between the finger and the conductor (which is at zero potential) gives rise to an electric field across the gap which on reaching the breakdown value for air precipitates a spark discharge.

One of the fuels used in a study of the nature and incendiary behaviour of spark discharges from the body done by Wilson (8) was a stoichiometric mixture of natural gas (methane) and air. The operator, while standing without his footwear on a sheet of insulating plastic, was charged to a known potential by his touching the output of a variable high voltage DC supply. The mixture was ignited by passing sparks from the finger to a grounded steel electrode of diameter 12 mm which was mounted inside an ignition chamber through which the gas mixture was passed. Similar sparks were passed to the electrode when it was connected to ground via a large capacitor which was in parallel with a high speed storage oscilloscope. The charge-time profile of a spark to the electrode, determined from the oscilloscope traces, showed that a spark discharge from the body consisted of several discrete sparks the first of which was the largest and alone likely to cause an ignition. The results also showed that for the gas to be ignited by a spark to the grounded electrode the energy of the charge stored on the body needed to be considerably greater than the MIE of the mixture. Further tests were done to determine the electrical resistances of several people by measuring the current passing through the body, under a known potential, via the ball of the finger when the latter was touching the grounded electrode. Values of the order of 10 kΩ were produced. This resistance was, of course, part of the discharge circuit.

From these results it was concluded that the discrepancy

between the energy of the charge on the body required to cause an ignition by a spark from the body and the MIE of the gas was largely a result of the fragmentation of the energy into discrete sparks and the absorption of a large proportion of it by the body resistance as heat.

Ignition tests with spark discharges from the body by Tolson (9) and Wilson (10) have shown that the energy of the charge on the body necessary for an ignition falls with the capacitance between the body and ground, but that the potential on the body stays almost constant. They also showed that the larger the electrode to which the spark was passed the greater was the potential needed on the body to cause an ignition. An exception to this latter behaviour occurred when the electrode was small enough to cause a corona discharge (10). The critical body voltages for the ignition of a stoichiometric mixture of methane and air (10) by sparks from the finger were 6.0 kV and 11.0 kV when passed to electrodes of diameter 1.0 mm and 12.0 mm, respectively. Taking the lower voltage, the corresponding energy of the charge on the body was 1.7 mJ. This is 4.4 times the MIE of the gas mixture. Tolson (9), using the most easily ignitable mixture of methane and air, caused an ignition with a spark from a freshly cut copper wire of 1.5 mm diameter when the potential on the body was 5.0 kV. The wire was connected to a metal rod which was held in the hand. At a potential of 5.0 kV, the energy of the charge on the body was 1.1 mJ. This is 3.9 times the MIE of the gas mixture. Further work (11) in which the most easily ignitable mixture of methane and air was ignited by sparks from the finger to a grounded steel electrode of diameter 0.5 mm showed that the lowest potential on the body to produce an ignition was 5.5 kV. The corresponding energy of the charge on the body was 4.6 times the MIE of the gas mixture.

The basic conditions necessary for a gaseous discharge to cause a fire or an explosion are the presence of a fuel, oxygen and an ignition source which together constitute the danger triangle (see 1.2). In addition, the interaction of these factors has to be in a prescribed manner (see 1.2.4). For the case of spark discharges from people, the above results show that the fragmentation of the spark, the absorption of energy by the body resistance, the capacitance of the body and the size of the electrode to which the spark passes are additional factors which influence whether or not an ignition will occur when the body is isolated from ground.

It is to be expected that the greater the number of conditions which require to be satisfied for a spark from the body to cause an ignition, the smaller the chance of it happening.

#### 2.6.1.2 *Nuisance discharges*

The accumulation of static charge on the body can lead to uncomfortable shocks when a grounded or large conductor is touched. Most people these days are familiar with the shocks felt on touching filing cabinets and metal light switches after walking over a carpet, or a car door after alighting from a vehicle. Despite this, shocks to the body caused by static electricity are, generally, regarded as being no more than a nuisance.

The severity of the shock felt by different people on passing a spark from the body at a given potential varies from person to person, depending on their sensitivity. This means that the threshold value of the potential on the body at which people begin to feel uncomfortable shocks is rather subjective. In tests on 12 men and women Wilson (12) examined their reactions to shocks felt in terms of the discomfort and how far down the finger they thought the shock travelled. An uncomfortable shock was related to a shock felt half-way down the finger. The results showed that at a body potential of about 2.0 kV one of the 12 subjects complained of an uncomfortable shock. None felt shocks at 0.4 kV and one of the subjects felt a mild shock at a potential of 1.4 kV on the body.

In the light of the above results and bearing in mind that the only condition required for people to feel static shocks on touching conducting surfaces is to have a potential on the body at least equal to their threshold value for discomfort, it is hardly surprising that static shocks are a common feature of everyday life.

### 2.6.2 Static discharges from clothing fabrics

The British Standard (13) requirement for clothing which is safe for use in areas where flammable atmospheres are present differentiates between situations where the MIE of the atmosphere is either greater than or less than 0.2 mJ. In the former case clothing of any material may be worn providing the person is suitably grounded by the use of antistatic or conducting footwear and floors. The resistance between the person and ground through the footwear and floor should not exceed 100 MΩ, and where there is the risk of exposure to mains voltage it should not fall below 50 kΩ.

For the case where the MIE of the flammable atmosphere is equal to or less than 0.2 mJ, the conditions for safety are as follows. The resistance to ground of the person through the footwear and the floor should not exceed 1.0 MΩ and, additionally, the surface electrical resistivity of the clothing should not be greater than 50 GΩ. As there is no requirement for a lower limit of the resistance to ground, precautions to prevent exposure to mains voltage should be taken.

In effect, the safety requirements imply two things. First, that for atmospheres with MIEs greater than 0.2 mJ, any static discharge from the clothing will not cause an ignition and that only precautions against spark discharges from the body need be taken. Second, that in atmospheres with MIEs of 0.2 mJ or less, static discharges from the clothing may cause ignitions and that precautions against this possibility should also be taken.

The value of the surface resistivity given above is achievable in clothing of natural fibres, e.g. cotton and linen, providing the relative humidity of the atmosphere is high (say 65 per cent or more). For clothing of man-made fibres, e.g. polyester and nylon, treatment with a suitable antistatic agent is required. Unfortunately, most antistats become less efficient in drier conditions as they rely on moisture from the atmosphere for their effect. Also they are not durable to repeated washing or dry-cleaning. Because of this, permanent antistatic measures have been introduced in recent years which have been applied to apparel and industrial fabrics. They make use of conducting threads containing metal or carbon-treated fibres (and other conducting materials) which are woven or knitted into the fabric in stripe or grid patterns. The European Standard (14) on the electrostatic properties of protective clothing specifies the same value for the surface resistivity of electrically homogeneous fabrics as does the British Standard. It also gives specifications for inhomogeneous fabrics, e.g. those with grids of conducting threads in which the conductor is at the surface of the threads. The requirements are that the surface resistances of the fabric should not exceed 1.0 GΩ, the grid must have a maximum spacing of 10 mm between the threads and that the grid should be grounded either through the body or directly.

It is noted in the Standard (14) that certain fabrics, e.g. those containing conducting core fibres, cannot be reliably assessed in terms of their surface resistivities. In such cases their electrostatic behaviour can be determined by spark discharge tests.

## 2.7 Literature


(1) Helmholtz, H. (1879). *Ann. Physik und Chemie*, **7**, 337.
(2) Klinkenberg, A. and van der Minne, J.L. (1958). *Electrostatics in the petroleum industry.* Elsevier, Amsterdam.
(3) Coehn, A. (1898). *Ann. Physik*, **64**, 217.
(4) Montgomery, D.J. (1959). Static electrification in solids. *Solid State Physics*, **9**, 139–197.
(5) Unger, B.A. (1981). Electrostatic discharge failures of semiconductor devices. *IEEE Proc. IRPS*, 193–199.
(6) Henry, P.S.H. (1953). The role of asymmetric rubbing in the generation of static electricity. *Brit. J. Appl. Phys*, **4**, Supplement No. 2. 531.
(7) DIN/VDE 0303 Teil 8/4.75, VDE-Verlag. Berlin, Germany.
(8) Wilson, N. (1979). The nature and incendiary behaviour of spark discharges from the body. *IOP Conf. Ser. No. 48*, 73–83.
(9) Tolson, P. (1980). The stored energy needed to ignite methane by discharges from a charged person. *Journal of Electrostatics*, **8**, Nos 2 and 3, 289.
(10) Wilson, N. (1983). The ignition of natural gas by spark discharges from the body. *IOP Conf. Ser. No. 66*, 21–27.
(11) Wilson, N. (1986). Incendiary behaviour of electrostatic spark discharges from human beings. In: *Performance of Protective Clothing.* ASTM 900. R.L. Barker and G.C. Coletta, eds, American Society for Testing and Materials, Philadelphia, 593–603.
(12) Wilson, N. (1972). The static behaviour of carpets. *Textile Institute and Industry*, August, 235–239.
(13) BS 5958 (1991). Code of practice for control of undesirable static electricity. Part 2. Recommendations for particular industrial situations.
(14) EN 1149–1:1995. Protective clothing – Electrostatic properties. Part 1: Surface resistivity (test methods and requirements).

# Chapter 3
# Electrostatic discharges as sources of ignition

## 3.1 Definitions

Every static discharge marks the end of a situation in which there has been an accumulation of electrostatic charge by whatever means. Sometimes it is a dissipation of charge to ground through a conducting material, or it can be a more spectacular event which can be both seen (see 3.2) and heard (see 3.3.2). As in other areas of science in which various effects need to be categorized, these visible and audible discharges fall into a group classified as gas discharges. This includes all manifestations from the scarcely perceivable corona discharge, through various types of spark discharge, to the flash of lightning which is the most spectacular event of all. The purpose of this section is to analyse, phenomenologically, the interrelation between gas discharges and electric fields. Readers who are more interested in the mathematical and physical details of the subject are referred to the relevant literature (1, 2).

## 3.2 Mechanisms of gas discharges

It is inevitable that there will always be free electrons (negative charges) in the atmosphere produced, for example, by cosmic rays. In an electric field these electrons are caused to drift in a direction opposite to that of the field. Their speed depends on the field strength and is limited by the frequency of their elastic collisions with the molecules of gas in the air. As the field strength is increased the electrons gain speed until a critical value is reached when the collisions become inelastic. At this stage the gas molecules suffering collisions with electrons release other electrons, and themselves become positively charged ions. This effect, known as ionization, leads to an avalanche of charge carriers (electrons and positive ions) which move with or against the field depending on their polarity. This movement of charge constitutes an electric current of magnitude dependent on the number of charges and the speed with which they move. This process is known as a gas discharge. In a homogeneous electric field, gas discharges extend along the entire length of the field and are initiated when the field strength reaches that required to cause an electrical breakdown of the ambient gas, which is usually air. Under normal atmospheric conditions this is reached in a uniform electric field of about 3 MV/m.

In an inhomogeneous field gas discharges occur first at the strongest part of the field when it is sufficient to cause an avalanche effect. As field strength is synonymous with field concentration, high field strengths occur most readily at the surfaces of conductors of small radii when placed in the electric field. An electrical breakdown close to a pointed electrode is manifested by a faint glow. This glow indicates an electrical stimulation of atoms which on relaxing to their original state emit photons of wavelengths determined by the nature of the gas in which the discharge takes place. In air the colour of the glow is, typically, violet and red in accordance with the line spectra of nitrogen and oxygen.

## 3.3 Types of gas discharge

Gas discharges may be classified into two principal types.

- one electrode discharges (see 3.3.1 to 3.3.5.2)
- two electrode discharges (see 3.3.6).

Spark discharges take place only between two electrodes. However, most gas discharges are of the one electrode type. One electrode discharges can occur when a grounded electrode is placed in an electric field coming, say, from a charged plastic material or a cloud of charged particles. For two electrode discharges the required electric field is produced by applying a potential difference between the electrodes. If, in addition to the potential difference, either the capacitance or the charge in the system is known, the energy released in the discharge may be calculated (see 3.3.6). For all one electrode discharges it is not possible to determine the energy released directly (see 3.4.1).

### 3.3.1 Corona discharges

In Fig. 3.1 is shown a representation of the field distribution between a charged insulator and a grounded conducting needle. As the needle is moved towards the insulator a corona discharge is emitted when the field strength at the point reaches the breakdown value for air. This discharge is visible only to eyes which are adapted to the dark and appears as a faint blue-violet glow. The onset of the corona is indicated by a small current passing through a microammeter interposed between the needle and ground. This current continues to flow as long as the field at the

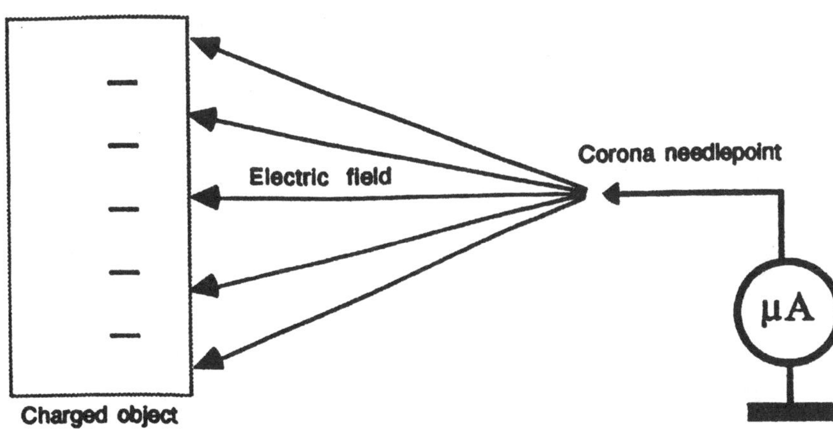


**Figure 3.1** Electric field between a charged object and a grounded needlepoint

needlepoint, from the charge on the insulator, remains at the breakdown value for air. The region of the breakdown is close to the needle and does not extend towards the insulator because of a rapid decrease in field strength in that direction. The charge carriers (electrons, ions), generated at the corona, move under the influence of the field, either to the insulator or to the needle depending on their polarity. In this way the air adjacent to the insulator becomes temporarily conducting and the positive charge reaching the insulator neutralizes a corresponding amount of negative charge. This effect can be demonstrated by a simple experiment.

Five triboelectrically charged polyvinyl chloride (PVC) tubes are placed on an inclined plane (Fig. 3.2). As they have the same polarity of charge they repel each other by means of electrostatic forces which are sufficient to prevent them from rolling down the plane under gravity thereby keeping them separated. When a grounded needlepoint is brought near to the tubes a corona discharge is initiated at the point which causes the charge on the tubes to become neutralized. Now, with only the gravitational forces acting on the tubes they roll down the plane and come to rest together. Similarly, a continuous corona current is produced when a grounded needlepoint is located near to the surface of a moving plastic film which is carrying a static charge. Any charge

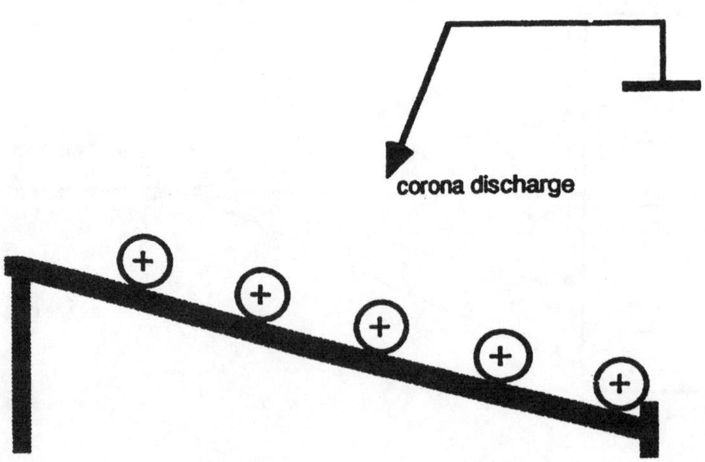


**Figure 3.2** Charged PVC tubes on an inclined plane

whose field is terminated at the needlepoint is neutralized by an equal and opposite charge coming from the corona discharge.

Many electrical processes are reversible, e.g. a motor-generator. The same applies to a corona discharge. In Fig. 3.1 the field at the point of the needle induces a high potential onto the needle which, at a critical value, leads to a corona discharge. Conversely, if in the absence of the field a high potential is applied to the needle, a corona discharge is produced. These corona discharges occur irrespective of the polarity of the charge on the object, or of the potential on the needle. To provide an idea of the effectiveness of needlepoints in initiating corona discharges some values of the potentials required on needles of different curvatures are shown below:

| Radius of curvature (μm) | 100 | 500 | 1000 |
|---|---|---|---|
| Corona onset potential (kV) | 2 | 4 | 6 |

It should be noted that corona discharges from needles with a negative polarity are initiated at rather lower potentials than are those from needles with a positive polarity. Corona discharges are classified according to the method used in generating them, as follows:

- active corona: needlepoint connected to a source of high potential
- passive corona: needlepoint connected to ground and exposed to an electric field.

Sometimes use is made of active corona to electrostatically charge objects or particles, e.g. copying machines and powder coating devices, etc.

#### 3.3.1.1 *Corona neutralizers for static elimination*

Unlike static charge on conductors, that on highly insulating materials cannot be rapidly dissipated by grounding the material. In such cases the procedure is to neutralize the charge on the insulator by depositing equal quantities of charge of opposite polarity onto its surface. This can be achieved by ionizing the air locally and relying on the field from the unwanted charge to draw an appropriate amount of charge to the surface.

Corona discharges are widely used for this purpose.

#### 3.3.1.1.1 *Active neutralizers*

This type of neutralizer usually consists of a number of sharp metal points which are mounted parallel to each other along the length of a rod. In some cases the rod is of metal so that the points are electrically interconnected. In others the rod is of an insulating material which ensures that the points are each electrically isolated from one another. Corona discharges are produced at the points by applying an AC high-voltage supply to the metal rod, or to the individual points in the insulating rod via suitable decoupling capacitors or resistors. The purpose of the latter is to limit the discharge from the points should they be touched or approach nearby conductors, thereby avoiding painful shocks to the body and possible ignitions in the presence of combustible atmospheres.

On switching on the AC supply the air molecules around the points become positively and negatively charged alternately in phase with the power supply, so enabling the neutralization of either positively or negatively charged insulators. Because there is a high rate of recombination of the positive and negative charges their lifetimes are only a few milliseconds. To avoid losing the charges by this means the points need to be located at a distance of not more than 20 mm from the charged surface. As negative coronas begin at rather lower potentials on the points than do positive ones, a surplus of negative charge has to be taken into account when using AC powered neutralizers.

The advantages of active neutralizers are that they can be switched on and off as required and can be easily located in positions of optimum efficiency. Their disadvantages are that they produce ozone, which needs to be controlled by venting, and that if there is a malfunction they may produce incendiary spark discharges in the presence of combustible atmospheres. Neutralizers made from conducting rods with points which are not decoupled from the supply by resistors or capacitors should not be used in places where combustible atmospheres are present.

#### 3.3.1.1.2 *Passive neutralizers*

The basic principles of passive neutralizers have already been described (see 3.3.1). Their advantages over active neutralizers are that they require no power supply, do not cause electric shocks when touched and present no danger of igniting combustible gases or vapours (see 3.4.2).

Passive neutralizers in the form of 'static combs' or metal wires wrapped with metallic tinsel are easily constructed and are relatively inexpensive. However, they too have their disadvantages. They are effective only as long as the electric field at their points does not fall below about 3 MV/m. As the unwanted charge is diminished by the corona discharges from the neutralizer, the field at the points is reduced to below the critical breakdown value resulting in a cessation of the discharges and the neutralization of the unwanted charge. This means that passive neutralizers are able only to reduce the unwanted charge from a high to a low level without eliminating it altogether. If this is all that is required then passive neutralizers are the best choice for the job.

## 3.3.2 Brush discharges

A similar representation to that for corona discharges (Fig. 3.1) is used for brush discharges (Fig. 3.3), the only difference being that the needlepoint is replaced by a spherical electrode. With the electrode in the same position as the needlepoint, relative to the charged object, no current is registered by the microammeter,

**Figure 3.3** Electric field between a charged object and a grounded sphere

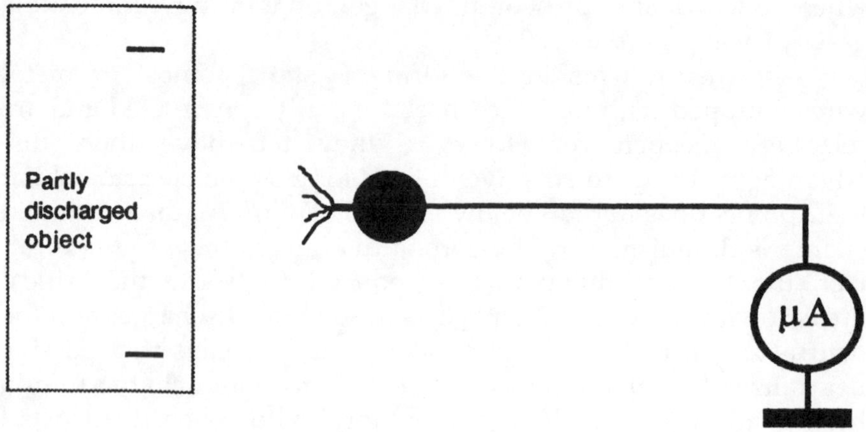


**Figure 3.4** Brush discharge

indicating the absence of a gas discharge. However, as the sphere approaches the charged object (Fig. 3.4) a brush-like discharge can be seen in the region of the sphere which is closest to the object. This appears as a short spark-like channel which starts at the sphere and fans out into faintly luminous filaments before disappearing in the gap between the electrode and the charged object. This 'brush' discharge also manifests itself by an audible 'crack' and a short pulse of current. As a result of the discharge some of the charge on the object, in the vicinity of the sphere, is neutralized.

On bringing a grounded spherical electrode near to the surface of a moving plastic film which is charged, a series of discrete brush discharges can occur at intervals depending on the speed of the film. The discharges begin when the electrical field strength at the sphere reaches the breakdown value for air. At this instant the distance between the plastic film and the sphere is much less than would be the case with a corona discharge. This means that the field strength in the gap is higher and has a greater spatial distribution than with a corona discharge which, in turn, leads to more ionization and a stronger current pulse.

Let us now consider another important factor to do with static discharges, namely, the magnetic field. Every electric current is surrounded by a magnetic field which, for spark discharges, has a constricting effect on the dimensions of the discharge channel. This can be demonstrated by a simple experiment. On passing a strong electric current along a thin-walled copper tube for a few milliseconds the tube suddenly collapses in on itself under the influence of the magnetic field. The same applies in the present case to the diffuse ion current starting at the sphere. If the magnetic field is strong enough the gaseous ions are compressed to a thin channel ('pinch effect') of high temperature plasma, while at the same time emitting a sharp 'crack'. As the charge on the object (Fig. 3.3) is neutralized so the electric and magnetic fields collapse. This instantaneous change in the magnitudes of the fields is accompanied by a release of energy in the form of electromagnetic waves which are radiated as radio-frequency signals and are detectable by means of an AM radio receiver. In general, brush discharges from highly charged surfaces are stronger when passed to larger electrodes. However, with lowly charged surfaces, brush discharges may only be possible to smaller electrodes which are capable of concentrating the field to the breakdown value for air. Usually, the larger the area of the charged surface, the stronger the brush discharge (3).

### 3.3.3 Super brush discharges

There is a kind of electrostatic discharge which does not fit into the category of the different 'one electrode' discharges already described in this chapter. It is a very intense type of discharge which takes place only in special circumstances. Up to now little interest has been shown in them as they very rarely occur in industry. An insight into these strange discharges can be seen by referring to Fig. 3.5. Five plastic tubes of the same material are electrostatically charged by rubbing them with a cat's fur. On placing them on top of each other, between insulating forks, the electrostatic force arising from the charge causes them to repel one another. These forces act against gravity, causing the tubes to separate with increasing distances between them the higher up the pile they are. Because of the method of charging, each tube has the same polarity of charge but with different quantities of charge on their surfaces. The charge on a tube and hence the charge density is limited by the maximum field strength of the surrounding atmosphere, as has already been mentioned (see 2.5.2

**Figure 3.5** Five like-charged plastic tubes in plastic forks

and 2.5.6). By bringing together, under gravity, the mutually repelling tubes the energy of the system is increased, thereby causing an increase in the charge density on the pile of tubes. It is under such conditions of heightened charge density that super brush discharges are possible (see 10.5).

By considering the coulomb energy and the charge on any system, the following can be deduced:

- Overcoming coulomb forces of attraction between opposite charges produces higher potentials (see Chapter 2).
- Overcoming coulomb forces of repulsion between like charges produces higher charge densities.

As each of the above processes causes an increase in the electrical energy, the energy transferred in any resulting gas discharge is also increased in both cases. This becomes evident when a grounded metal electrode is brought up to one side of the charged tubes. A more energetic brush discharge than is ordinarily the case is caused in which brush-like luminous channels are produced which sometimes reach the surfaces of the tubes. The reason for this is that the energy of the discharge is determined partly by the charge density at the surface of the insulator. A similar effect is reported by Glor (4) when charged granules are fed into a flexible intermediate bulk container (FIBC). It seems reasonable to assume that this kind of super brush discharge, as well as occurring on heaps of highly charged powder, might also be a link between ordinary brush discharges (see 3.3.2) and cone or bulking discharges (see 3.3.4).

### 3.3.4 Cone or bulking discharges

Cone or bulking discharges are sometimes visible and have been photographed at the surface of a heap of plastic granules being fed into a silo. In recent years a number of papers have been published on this phenomenon among which, of particular note, is that of Glor and Maurer (5). A photograph of cone discharges taken with a camera mounted on top of a silo, as the latter was being filled with plastic granules, is shown in Fig. 3.6.

The mechanism of cone discharges is in some ways similar to that of super brush discharges. When feeding charged particles

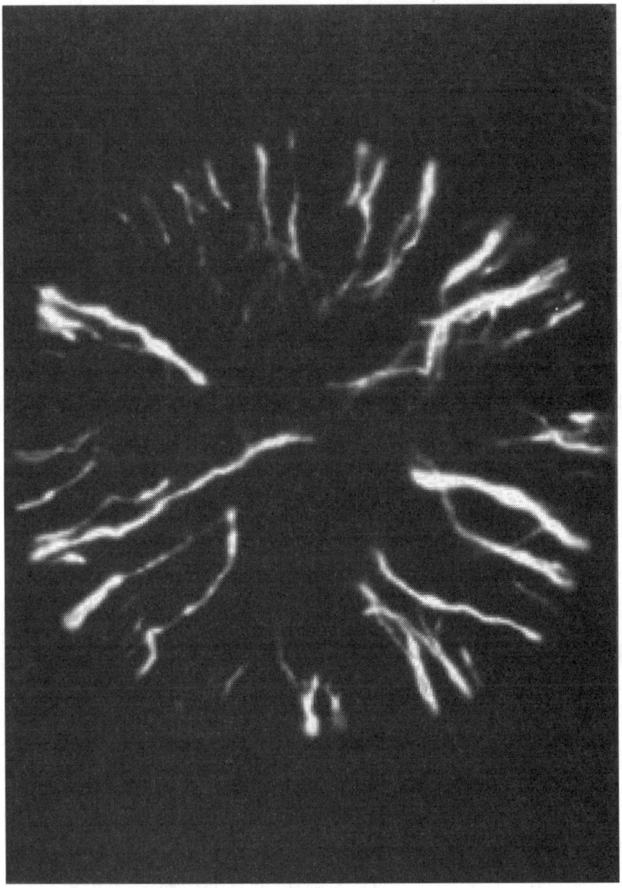

**Figure 3.6** Cone (bulking) discharges inside a silo

into a silo, there is an accumulation of charge at the heap. The field from this charge exerts repulsive forces on the similarly charged particles which are falling onto the heap. As the gravitational forces on the particles act against the repulsive forces there is an increase in the charge density of the bulk material at the heap. If charged particles continue to fall onto the heap after the field strength in that region has reached the breakdown value for air, gas discharges will always occur in a direction towards the conducting wall of the silo, which is grounded. But as the available energy has been increased so the discharges may extend over the entire radius of the silo. This is accompanied by a sharp drop in the electric field and it takes some time for sufficient charge to accumulate again on the heap to initiate fresh discharges. As parts of the discharge channels are strongly luminous it follows that the 'pinch effect', which has already been discussed, is taking place. Cone discharges can be detected as high frequency signals, as are brush discharges.

### 3.3.5 Propagating brush discharges

Compared with the types of gas discharge already described, propagating brush discharges (Lichtenberg discharges) are, with the exception only of lightning flashes, the most powerful kind of electrostatic discharge. As the inductive reactance (inductivity) of a propagating brush discharge circuit is extremely low in comparison with, say, that of a spark discharge circuit, the current rise-time of a propagating brush discharge is correspondingly much faster, resulting in sharp pulses. The effect of the electromagnetic radiation from such pulses can be to induce damaging potentials and charges on any electronic equipment in the vicinity.

#### 3.3.5.1 *Propagating brush discharges on shielded systems*

Under normal atmospheric conditions the maximum theoretical field strength at the surface of an insulating material (dielectric) is not more than 3 MV/m, corresponding to a charge density of about 26 $\mu C/m^2$. This limitation of the charge density arises because of the electrical breakdown of the air and the partial neutralization of charge at the surface by the ions so produced. In practice charge densities well below the theoretical value are usual. If the insulating material is in the form of a thin foil and is placed onto a grounded conducting plate the charge density on the upper surface of the foil may reach a much higher value than that quoted above. The reason for this is that as the electric field from the surface charge is directed mainly towards the plate, through the dielectric, the charge density is now limited only by the breakdown strength of the dielectric which, typically, is 100 to 1000 times higher than that of air. Such high charge densities can only be generated by spraying charge from a corona discharge (see 3.3.1) onto the surface and by certain industrial operations, e.g. the pneumatic transport of powders through insulating tubes and the winding up of charged insulating foil. They cannot be produced by simply separating or rubbing the surfaces.

Should the field in the dielectric, from the charge on its surface, reach the breakdown value of the dielectric, a spontaneous discharge occurs resulting in a puncture of the foil (Fig. 3.7). Starting from the puncture, a very high electric field is created parallel to the foil's surface which initiates a series of strong surface discharges, thereby releasing most of the surface charge. The discharges are themselves compressed into spark-like channels by the strong magnetic fields ('pinch effect'). As energies of up to 10 J may be released in a propagating brush discharge, the possibility of severe physiological shocks to personnel must be taken into consideration. In assessing the occurrence of propagating brush discharges on insulating layers resting on conducting backings, Maurer *et al.* (6) have revealed that such discharges do not occur when the thickness of the insulator is more than 10 mm or when the breakdown potential of the insulator is less than 4 kV.

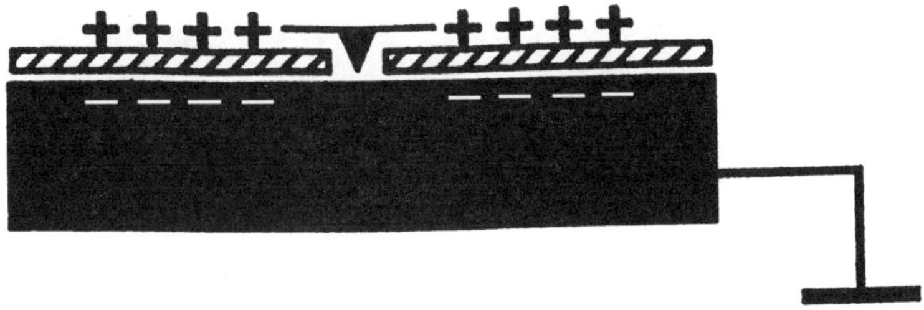

**Figure 3.7** Propagating brush discharge (shielded system)

#### 3.3.5.2 *Propagating brush discharges on unshielded systems*

Propagating brush discharges can also occur on the surfaces of free-standing insulating materials, e.g. plastic bags and plastic tubes, when bipolar charges of sufficient density are produced on their surfaces (Fig. 3.8). As with shielded systems such discharges are only possible subject to the conditions mentioned above. This type of discharge may sometimes be observed with FIBCs made from insulating materials, e.g. polypropylene, on filling them with highly charged materials. When such a bag is left standing for some time charges of opposite polarity to that on the contents of the bag are attracted to the outside surface. As the field from this charge is directed into the bag, very high charge densities can be built up on the surface. If the surface is then touched by a grounded or large conductor, or a person, a propagating brush discharge can be initiated.

**Figure 3.8** Propagating brush discharge (unshielded system)

### 3.3.6 Spark discharges

Discharges between flat parallel metallic electrodes are usually initiated at the edges of the electrodes where the charge density is greatest. By bending the edges of the electrodes into a curved shape (Rogowski profile) the field between the electrodes is made more uniform, thereby avoiding premature discharges at the edges. The discharge occurs when the field strength between the electrodes reaches the breakdown value for air and a characteristic of the discharge channel is that it extends the entire distance between the electrodes. Knowing the potential difference $U$ across the electrodes the corresponding field strength $E$ is given by,

$E = U/d$ ($d$ = distance between the electrodes)

For the case of a spark discharge from a charged capacitor (Fig. 3.9) the energy $W$ released is calculated using the equation,

$W = 0.5CU^2$ where $C$ is the capacitance.

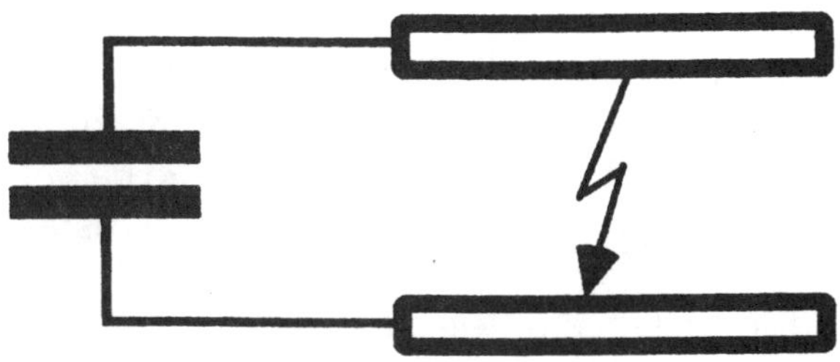

**Figure 3.9** Spark discharging of a capacitor

Spark discharges, for instance, may occur between an isolated metal drum, which has become charged during filling, and a nearby large or grounded conductor. The drum behaves like a charged capacitor. As the human body is electrically conducting, it also acts as a capacitor when isolated from earth and is capable of delivering spark discharges.

If the capacitance and the potential of an isolated conducting system can be measured, e.g. a metal drum or flange, or the human body, then the energy available for a spark discharge from the system can be calculated.

## 3.4 Incendivity of gas discharges

For the ignition of any fuel (gas, vapour or dispersed solid) the igniting source needs to have a temperature which is at least equal to the igniting temperature of the material. In general all gas discharges in which the 'pinch effect' occurs are so hot that they may ignite any fuel providing there is sufficient energy. Corona discharges, on the other hand, being of a diffuse nature are not capable of causing ignitions because they show no 'pinch effect', i.e. they are 'cold'.

### 3.4.1 Assessment of ignition energy

In order to find out if an ignition can occur it seems logical to compare the MIE of the material with the energy released in the static discharge. Although in the assessment of safety this criterion is used worldwide there are other factors which have to be considered.

Energy is a measure of the capacity for doing work, but how the energy is expended temporally and spacially is of crucial importance. To determine the true MIE of a material the duration and spacial distribution of the igniting spark have to be optimal. Therefore, the potential across the electrodes, the distance between them, the shape and size of the electrodes, and the capacitance, resistance and inductance of the discharge circuit have to be carefully selected. Usually the circuit is designed to practically eliminate the effects of resistance and induction, and the remaining parameters are adjusted for optimum conditions. In this way the MIEs of a large number of materials have been established for use in safety investigations. But still there remain doubts about the significance of MIE for the case of non-spark-like discharges. In 1965 Gibson and Lloyd (7) introduced the term 'equivalent energy' for discharges from insulating surfaces which is defined as follows. A discharge possesses an equivalent energy of, say, *X* joule if it is just capable of igniting an explosible gas/air or vapour/air mixture with a MIE of *X* joule. Tests showed that brush discharges may be assigned equivalent ignition energies in the range of 1–4 mJ. This range includes values of established MIEs for very sensitive dusts. However, to date there is no evidence of dusts having been ignited by brush discharges and so it is still questionable as to whether or not 'equivalent energy' provides a correct assessment of non-spark-like discharges. Schwenzfeuer and Glor (8) are working on this problem but, as yet, no definite results are available.

As the igniting power of certain types of one electrode gas discharges is at present a subject of controversy among experts we do not wish to prejudice the issue here. Statements concerning the incendiary behaviour of discharges are made in as far as they are necessary for a proper understanding of the case histories. This means that with certain case histories, especially those involving brush discharges, it is sometimes not sufficient to view the evidence from only one angle.

### 3.4.2 Table of incendivity

Table 3.1 represents a summary of Chapter 3. It applies only to fuel/air mixtures under normal atmospheric conditions and indicates the likelihood of ignitions which might be caused by different types of gas discharge.

**Table 3.1** Incendiary behaviour of different fuels by gas discharges

| Type of gas discharge | Gases and vapours MIE: 0.2→1.0 mJ | Sensitive dusts MIE: 1.0→10 mJ | Dusts MIE > 10 mJ |
|---|---|---|---|
| Corona | no | no | no |
| Brush | possible | not proven | no |
| Super brush | possible | suspected | suspected |
| Cone | possible | suspected | suspected |
| Propagating brush | yes | yes | yes |
| Spark | calculable | calculable | calculable |

## 3.5 Traces left by gas discharges

In searching for the causes of damage by fire, investigators often find residual traces which can be useful in identifying the source of the ignition, e.g. the coloration of metal shafts caused by overheated bearings or the smouldering of electrical equipment. Such traces are relatively easy to identify, but what about the detection of those left by electrostatic discharges?

From experience it is a known fact that evidence of such discharges is identifiable, but because the energy released is often very small the traces left behind can be correspondingly tiny. A list of a number of examples of the different kind of marks which may be caused by various types of gas discharge is given below.

- Corona discharges are events which are extended in time, as has already been explained (see 3.3.1). They release energy at a slow rate relative to those of other gaseous discharges and do not leave any definite traces as such. However, they can produce secondary effects, e.g. an improvement in the wettability of polymer surfaces to which they have been directed.
- Brush discharges from a charged insulator leave behind fern-like traces on the surface which, usually, are not immediately visible. In order to see them they have to be 'developed' by applying, for example, Xerox toner which works straight away. However, airborne dust particles that are attracted to the surface of an insulator, following a brush discharge from its surface, will also reveal the traces after a period of time.

- Super brush discharges display a similar pattern on surfaces from which they are released to those of brush discharges but on a much larger scale.
- Cone discharges leave no traces on the powder heaps from which they originate. However, at those places on the silo walls where the discharges are terminated, tiny erosion marks are sometimes found when viewed under a magnifying glass.
- Propagating brush discharges are extremely powerful and can leave branch-like traces covering large areas of an insulating surface. They are revealed in the same manner as are brush discharges and can be so powerful that their route may sometimes be identified instantly as traces of molten plastic on the surface. At the start of a branch a puncture or crater-like perforation can often be seen. When a propagating brush discharge is initiated by a grounded electrode approaching the surface, erosion marks can be seen on the electrode.
- Spark discharges can produce erosion and even melting at the electrodes across which they pass, depending on the amount of energy released.

As has been described above electrostatic discharges may leave traces which, now and again, are very difficult to find. They furnish only qualitative evidence of gas discharges and, therefore, are not necessarily a means for drawing reliable conclusions about the incendiary behaviour of the discharges. Proof of an electrostatic ignition can only be achieved by reconstructing the conditions of the accident. It is in this manner that many of the case histories to follow are dealt with.

## 3.6 Literature

(1) Taylor, D.M. and Secker, P.E. (1994). *Industrial Electrostatics.* Research Studies Press Ltd, GB.

(2) Luttgens, G. and Glor, M. (1989). *Understanding and Controlling Static Electricity.* expert-verlag. (Also published in German (1993) by expert-verlag under the title, *Elektrostatische Aufladungen begreifen und sicher beherrschen.*)

(3) Richtlinien 'Statische Elektrizitat' ZH1/200 (1989). Hauptverband der gewerblichen Berufsgenossenschaften, Lindenstr. 78–80, D-53757, Sankt Augustin.

(4) Glor, M. (1988). *Electrostatic Hazards in Powder Handling.* Research Studies Press Ltd, GB.
(5) Glor, M. and Maurer, B. (1994). Zundfahigkeit von Schuttkegelentladungen. BMFT-FB (13 RG 9002). D.
(6) Maurer, B., Glor, M., Luttgens, G. and Post, L. (1987). Hazards associated with propagating brush discharges on flexible intermediate bulk containers, compounds and coated materials. *Inst. Phys. Conf. Ser.*, No. 85, 217–222.
(7) Gibson, N. and Lloyd, F.C. (1965). The incendivity of discharges from electrostatically charged plastics. *Brit. J. Appl. Phys.*, **16**, 1619–1631.
(8) Schwenzfeuer, K. and Glor, M. (1995). Ignition tests with brush discharges and dusts. *Inst. Phys. Conf. Ser.*, No. 143, 125–128.

# Chapter 4
# Principles of safety

Having reached this stage in the book the reader may well be wondering about the factors that need to be considered in overcoming problems arising from static electricity and the principles on which safety precautions are based. From the systematic investigation of the case histories described in the following chapters various requirements for safety and the methods used in their implementation have been dealt with.

Although static charges are almost everywhere, they are, for the most part, benign in terms of the danger and nuisance they can cause because of their small quantity. It is when charges of different polarities become separated and then accumulate in more substantial quantities that problems can arise.

Charge separation can occur in a number of ways but the most frequent means are by the contact and separation of materials and by electrostatic induction. In Fig. 4.1 is shown a schematic diagram of the development of different types of incendiary gas discharge arising from accumulated static charges.

When the materials concerned are of a conductive (or dissipative) nature the residual charges on their surfaces after separation are minimal, if not zero, and present no danger. However, when at least one of the materials is an insulator it will become charged and will retain the charge for a more or less long period of time. The other material, whether it is a conductor or an insulator, will acquire an equal and opposite charge to that on the insulator.

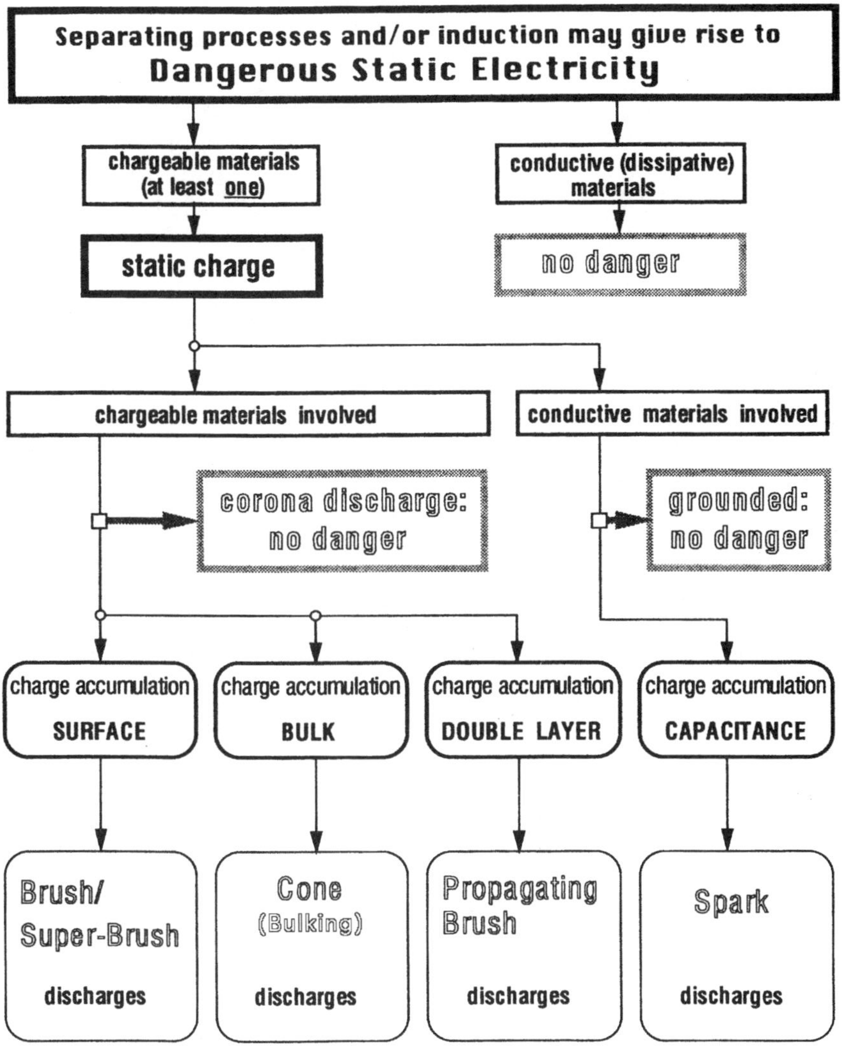


**Figure 4.1** Development of different types of incendiary gas discharge

Should the charges on these surfaces be large enough, gaseous discharges can occur which, in the presence of a combustible atmosphere, might cause an ignition. It is, therefore, important to know the procedures that should be followed to avoid such discharges from insulating materials and from conductors.

Depending on the circumstances, the grounding of conductors can be an easy means of overcoming a potential danger providing it is done reliably and permanently. However, should such materials not be grounded they will behave as capacitors and when charged will be capable of releasing spark discharges. Grounding a charged insulator would be practically ineffective in getting rid of the charge because the latter has a low mobility. However, a charged insulator may, again depending on the circumstances, be partially or wholly neutralized by means of a corona discharge. Such discharges operate by depositing charge on the insulator of opposite polarity to that which is already there. When the charges are equal in magnitude the charge on the insulator is, effectively, neutralized. Even the partial neutralization of charge by this means is usually sufficient in preventing incendiary gas discharges from the insulator.

The accumulation of charge on an insulator may be on the surface, within the bulk, or in the form of a double layer. Charge accumulated on surfaces can lead to brush and super brush discharges. Charge within the bulk may give rise to cone discharges and charge accumulated in the form of a double layer can produce propagating brush discharges. The incendiary behaviour of these different types of discharge, as well as of spark discharges, is given in 3.4.2 (Table 3.1).

As would be expected, the size of a charged material is one of the factors which determine the quantity of charge that it can hold. It follows, therefore, that the probability of an incendiary discharge falls as the area of the insulating surface decreases. Brush discharges from insulating surfaces of area less than 2000 $mm^2$ are not able to cause the ignition of sensitive mixtures of hydrocarbon vapours and gases with air. For cone discharges to be incendive the volume of the charged material needs to be, at least, a few cubic metres (see 3.6, (5)). Double layers of charge on continuous sheets of insulating plastic are safe from causing propagating brush discharges providing the sheets are of thicknesses greater than 10 mm or their breakdown potentials are less than 4.0 kV (see 3.6, (6)). Spark discharges from small conductors of capacitance below 3 pF are not able to ignite hydrocarbon vapours and gases (see 3.6, (3)).

As the safety procedures referred to above are those applied to many of the case histories to follow, it might be tempting, when faced with a safety problem in a plant, to identify it with one of the case histories and act accordingly. However, the reader should beware of this possible trap as the facts of the problem may differ slightly from the case history, thereby requiring a different approach to its solution. It could also lead to the imposition of restrictions which are not necessary.

Studying the case histories will afford a useful training in becoming aware of the dangers that can arise from static electricity. However, in addition to this knowledge it is advisable always to consult the relevant guidelines as they appear in British Standards, German Berufsgenossenschaft and the American NFPA. As most guides are usually several years old it is worth noting that a CENELEC report entitled, 'Safety of machinery – Avoidance of hazards arising from static electricity' is at present in preparation. The final form of the work is, as yet, unknown but undoubtedly it will become one of the most comprehensive documents on electrostatic hazards.

# Chapter 5
# Case histories related to brush discharges

## 5.1 Ignition in a heated tank containing diphenyl

Diphenyl was being stored in a cylindrical metal tank of volume 100 $m^3$ (Fig. 5.1) which was thermally insulated on its outer wall. Inside the tank and parallel with its axis were two heating pipes through which water flowed at a temperature of 90°C. The temperature at which diphenyl becomes solid is 69°C and its flash point is 113°C. Because of a change in the production process it was necessary to empty the tank completely and clean it out. On refilling the tank an explosion occurred when the tank was about half full. The effect of the explosion was to blow away the dome of the tank and cause a large fire.

On investigating the accident we were able to gather enough evidence to show fairly conclusively how it had come about. The temperature of the liquid diphenyl, as it was being delivered via a pipe, was about 120°C and as air had entered the tank during cleaning a combustible mixture of diphenyl vapour and air could have been produced. The volume resistivity of liquid diphenyl is about 100 M$\Omega$ m which, from experience, is known to be the threshold above which the liquid cannot easily dissipate any static charge on it. Further, the accumulation of charge on the liquid as it passes through the pipe depends on its rate of flow. The faster

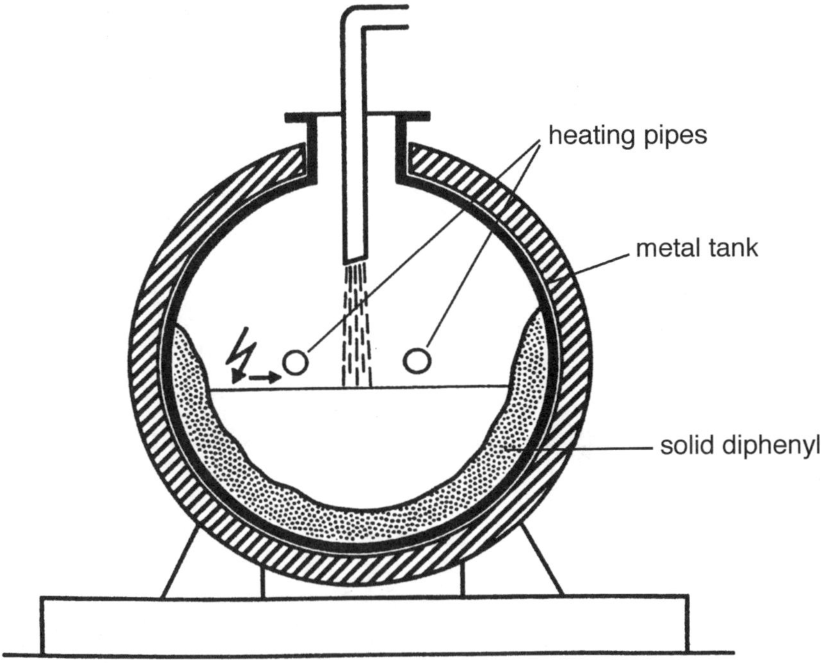


**Figure 5.1** Cross-section of a cylindrical heating tank

the movement of the liquid the more charge it will retain. As the tank was cold at the start, some of the liquid diphenyl had become solid on the inner wall. It is known that the effect of solidification on the volume resistivity of diphenyl is to increase it by more than four orders of magnitude, as can be seen by referring to Fig. 5.2. Thus, an electrically insulating layer was formed on the inside surface of the tank which would have prevented the charge in the liquid from escaping quickly. The heating pipes would only become effective when they were covered with liquid, i.e. when the tank was at least half full. As the level of the charged liquid approached the heating pipes it is likely that a brush discharge occurred between the pipes and the surface of the liquid which ignited the vapour/air mixture.

In general, when brush discharges are the suspected ignition source, it is difficult to evaluate the probability of an ignition. However, from the accumulated evidence the indications were, in

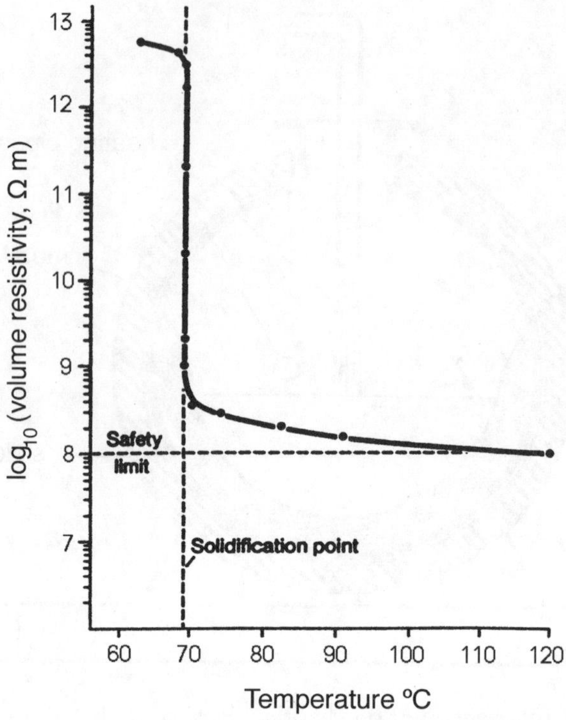


**Figure 5.2** Volume resistivity of diphenyl related to temperature

this case, that a brush discharge was the only possible cause of the accident.

*Suggested safety measures*

- The recommended procedure for the avoidance of the formation of a combustible vapour/air atmosphere would be to purge the tank with nitrogen or carbon dioxide.

## 5.2 Pouring flaked product into an agitator vessel

One hundred and twenty polyethylene (PE) bags of organic flake were to be fed into a $10\,m^3$ stainless steel agitator vessel. Beforehand the agitator vessel was rinsed with water, drained while purged under nitrogen pressure, and dried by heating. The mouth of the vessel was then closed to keep the vessel clean and free of oxygen. This was done on the Friday afternoon; on the following Monday morning workers began to feed the flakes into the vessel.

While emptying the 86th bag, at a stage when the vessel was about half full, a tongue of flame suddenly shot out of the mouth of the vessel causing a worker to be burned, after which the flame went out. Fortunately his eyes and hair escaped burning as he was wearing protective goggles and a helmet. The flame caused no blackening of either the vessel or the product but the PE bag and the helmet were partially shrivelled by the heat.

The management were immediately of the opinion that such an odd occurrence could only have been caused by an electrostatic discharge. As a result we began a thorough investigation of the accident which included an examination of all the connections to the vessel. A pipe was found which had at one time been used to feed ethylene oxide into the vessel but which by then had been standing idle for a long time. It had two closed shut-off valves, connected in series with each other, which had been tested under water pressure and shown to be functioning properly. Test statements had been noted. The shut-off valves were tested twice again with water at a pressure of 5 bar (0.5 MPa) and shown to be perfectly tight. However, on testing the valves using ethylene oxide at a much lower pressure, a small leakage was found. This immediately threw suspicion onto the ethylene oxide, particularly as when it burns no soot is produced. Knowing about the leakage we were able to reconstruct the course of events conclusively. During the long standstill at the weekend ethylene oxide was able to seep slowly into the vessel. Being denser than nitrogen it accumulated at the bottom of the vessel thereby displacing nitrogen to the outside through the mouth of the vessel which had not been tightly closed. On pouring the product into the vessel ethylene oxide was displaced from the bottom up the sloping sides of the product until it reached the opening at the top. There it could mix with air to produce a combustible atmosphere which was ignited by a brush discharge from the PE bag (MIE of ethylene oxide = 0.07 mJ).

*Suggested safety measures*

- Disconnect all pipes from the equipment that are not necessary.
- If their disconnection is not possible check the valves under gas pressure, not under water pressure.

- After long periods of idleness purge the vessel with nitrogen before use.

## 5.3 Filling pipe blocked with sulphur leading to the ignition of methanol

Pure finely ground sulphur was being poured out of paper bags into a 6 m long filling pipe leading to an agitator vessel containing a mixture of methanol and water (flash point 13°C). This procedure had been carried out about 400 times when, on this occasion, an explosion occurred which required an immediate investigation. According to an eye-witness the explosion had happened about ten seconds after the last paper bag was emptied. During the investigations it was found that the exhaust system on the vessel was very weak because of a congested suction device. As a result of this a combustible methanol/air mixture was able to move up into the pipe.

It is well known that sulphur can become highly charged electrostatically when being moved about but despite this no one had thought that such charging could lead to an incendiary brush discharge.

All chemical reactions were excluded as possible sources of ignition. What provided the clue to solving the problem was the time delay between emptying the last bag and the explosion. On questioning the workers it was revealed that every now and again the fine sulphur would block the 300 mm diameter steel pipe. Armed with this information we carried out the following experiment. A slide valve mounted near to the lower end of the pipe was closed and a bag of sulphur was emptied into the pipe. When the valve was opened it was found that the sulphur had become stuck in the pipe. On striking the pipe the sulphur was released, giving rise to an electric field of strength 600 kV/m and a strong radio-frequency signal which was detected by an AM receiver (see 3.3.2). The test was repeated twice with the same result, thereby establishing the possibility that a brush discharge from the charged sulphur could have ignited the methanol/air mixture in the filling pipe, with the flame then being propagated into the vessel.

*Suggested safety measures*

- Ensure that the exhaust system is regularly maintained.
- In the presence of possible ignition sources explosible atmospheres should be purged with nitrogen.

## 5.4 PE liner slipping out of a paper bag

An inorganic powder was emptied out of a paper bag, containing a PE liner, into an agitator vessel in which was a mixture of benzene and methanol at a temperature of 22°C. As soon as the bag was empty flames appeared at the opening of the vessel. Because all other ignition sources, especially unintentional chemical reactions, could be reliably excluded as the cause of the accident, only static electricity needed to be taken into consideration. Spark discharges from the worker could also be excluded as he was wearing conducting shoes and was standing on a conducting floor. Furthermore, there were no other isolated conductors in the vicinity. It was, therefore, felt likely that a brush discharge from the charged PE liner was the cause of the ignition.

Paper bags with PE liners do not usually give rise to hazardous static electrification but in this case the PE liner was sewn to the paper bag at its upper edge. Depending on the way the worker held the paper bag when emptying its contents into the vessel the PE liner would either have stayed inside the bag, or slid out of it with the rest of the product. It would not, of course, have fallen into the vessel because of its attachment to the paper bag. In order to understand the electrostatic effects it is important to know whether or not the PE liner stayed in the bag. Should it have remained inside the bag the charge on it would have been shielded by a similar charge of opposite polarity induced onto the bag. However, if the PE bag had slipped out of the paper bag the shielding effect would have been greatly reduced and the charge on the former would have been free to cause a brush discharge which could have ignited the solvent/air mixture.

*Suggested safety measures*

- When using paper bags containing PE liners in the presence of combustible vapour-enriched atmospheres make sure that the liner remains inside the paper bag.

## 5.5 Ignition caused by an antistatic PE bag

When emptying product from PE bags into vessels in which explosible gaseous atmospheres are present, one has always to take into account the possibility that incendiary brush discharges may occur. If it is essential to use PE as the packaging material then it should be treated either with a suitable antistatic agent, or glued to paper as a means of providing a shielding effect. In the present case the former was preferred. During the production of the PE used for making the bags an antistatic agent was added which migrated to the surface of the film thereby reducing its electrical resistance. According to the German Guidelines (see 3.6, (3)) the surface resistivity of the PE has to be less than $10\,G\Omega$, a value that could readily be achieved. The bags made from this material were labelled 'antistatic packaging'.

We learned about an explosion which occurred at a customer's plant when a pigment powder was being emptied out of an antistatic PE bag into a reactor vessel which contained some methanol. Those involved were amazed that this should have happened and quickly called us in to begin a very thorough investigation of the accident.

Having learned that the bags were made from an antistatic PE we, at first, ruled them out as the source of an electrostatic discharge and suspected that an exothermic chemical reaction had occurred. However, as it was a simple solvent process that was being carried out at room temperature this idea was not pursued. Further, as the worker concerned was wearing conducting shoes and was standing on a metal grating the possibility of an incendiary spark discharge from the body was dismissed. Suspicion, therefore, fell on the PE bag. Perhaps it had been mislabelled and was not one containing the antistatic agent. In fact, determinations of the surface resistivities of the remains of the suspected bag and of other still filled bags showed values of about $10\,T\Omega$! These results indicated that the blame for the accident might have lain with the manufacturer for being negligent in the production of the antistatic bags and that we could now put the case on file. However, he was adamant that nothing had been changed on the production line and that he could reliably prove this by random tests on his bags. This led us to suspect that the effect of the antistatic agent might be decreasing with time and so we checked this by a series of tests in which the surface resistivities of freely hanging bags were determined at selected times over a period of six months. The values were found to be always within the expected range of one order of magnitude. At this stage we were beginning to think that there was no way of solving the problem unless there was some form of chemical reaction between the pigment powder and the antistatic agent! This idea was rejected by the chemists, as neither the antistat nor the pigment were highly reactive. We then thought that there might be a tripartite interaction between the antistat, the pigment and the PE. On further study it was found that the antistat was not compatible with the polymer as it too readily migrated to the surface, although, because of its low vapour pressure, the antistat would normally remain at the surface for a long time. However, when a material of low particle size makes contact with a layer of antistat a very large surface area of the material is presented to the latter. The result, in this case, was that sufficient of the antistat diffused into the pigment to cause the resistivity of the PE bags to rise to the high value quoted above. Indeed, tests have shown that some very fine pigment powders can absorb the entire antistatic content of treated PE within a few days thereby robbing the latter of its antistatic properties.

*Suggested safety measures*

- The antistatic treatment of PE bags is not always permanent and should be evaluated in individual cases.

## 5.6 Impregnation of a glass fibre fabric

Following the occurrence of accidental fires in processing plants experts in electrostatics are sometimes asked the question why, after years of using the same method of operation, does an ignition suddenly occur. One can always give the general reply that electrostatic discharges are, for the most part, very weak and therefore capable only of igniting optimally combustible atmospheres (see 1.3.3). But, of course, it would be folly to rely on this fact alone without checking all the other possibilities.

In the present case a fleece of glass fibre was being impregnated with an epoxy resin dissolved in methylethyl ketone (flash point −4°C). The untreated fleece was unwound from a roll, passed through a heated duct for drying and on to an impregnation chamber where it was immersed in a dip tank containing the resin. It was then passed across a warm-air blower as a means of polymerizing the resin.

Immediately prior to the accident the fleece had become ruptured in the impregnation chamber, a problem which occurred almost daily. When this happened the dip tank was lowered out of the way so that the return duct and the fleece were no longer immersed in the epoxy resin. On pulling the untreated fleece through the return duct to reconnect it to the torn end a blast flame occurred which ignited the mixture in the dip tank causing a large fire and the destruction of the entire chamber.

Now the question arose as to what, on this occasion, had been done that was different to the everyday routine. The electrostatics expert was particularly interested in the effect of drying the fleece on its electrostatic properties. This was an important step in the treatment of the fleece because, as with all polymer processing, the presence of too much moisture is undesirable. By varying the temperature of the heating duct attempts had been made to control the levels of moisture in the fleeces to an optimum value. The batch of fleece preceding that at the time of the accident contained an unusually high amount of moisture requiring the operator to increase the temperature of the heating duct to 90°C. The fleece that was being processed at the time of the accident had a much lower moisture content but the temperature of the duct had not been reduced accordingly. The operator possibly thought that the drier the fleece the better would be the quality of the resin treatment! The effect, as later tests showed, was to raise the surface resistivity of the fleece from its usual value of 1 TΩ to 100 TΩ.

Experience shows that no significant electrostatic charges are to be expected at resistivities below 1 TΩ for processing speeds of 1 m/s or less. However, at 100 TΩ high charge densities are possible which may cause incendiary gas discharges.

*Suggested safety measures*

- At processing speeds of up to 1 m/s the surface resistivity of the product should not exceed 1 TΩ in the presence of combustible vapour atmospheres.

## 5.7 Shaking fine dust out of a PE bag

As has already been pointed out in 3.4.1 brush discharges, in the opinion of experts, are not capable of igniting dusts but can ignite combustible solvent vapours. Accordingly, it was permitted to pour combustible powder out of PE bags into an empty 5001 vessel which was partially covered with a lid (Fig. 5.3). All the safety measures to avoid elecrostatic sparks had, of course, been taken. There was a conducting floor, the worker had conducting shoes and gloves and all metal parts were grounded.

**Figure 5.3** Shaking powder out of a PE bag

A few months later we received a message that an explosion had occurred at the filling place and that a worker had been severely burned by a tongue of flame. As nothing had been changed since the accident it was possible to demonstrate clearly that the ignition occurred just when the worker had finished shaking the dust out of the PE bag. However, in view of the fact that the required safety measures were in place at the time, the question arose as to whether or not brush discharges could, indeed, ignite dusts. As we wanted to find out if the vessel had been empty before filling it with powder we questioned the staff. We were told that as they were dealing with pharmaceutical products everything had to be perfectly clean and that the vessel had definitely been empty. On being asked how they cleaned the vessel we were told that ethanol was used. This, of course, raised the possibility of ethanol vapour having been present in the vessel. We learned that after the vessel had been emptied ethanol was sprayed onto its inner walls. The ethanol was drained away through an outlet valve which afterwards was closed ready for the vessel to be refilled. As the walls would be wet with ethanol an ethanol vapour/air mixture would be present in the vessel of a concentration dependent on the time the covering lid was kept open. This could have been within the explosion range of the mixture, and would also have depended on the amount of air introduced into the vessel as the PE bag was being shaken.

On that fateful day a brush discharge and an optimally combustible ethanol/air mixture could have coincided thereby causing an ignition of the vapour, a swirling up of the powder and the ignition of the latter.

*Suggested safety measures*

- Investigate thoroughly, not only the process in question, but also the possible consequences of preceding ones.
- Do not use chargeable PE bags in the presence of combustible solvent vapours.

## 5.8 Ion exchanger resin in toluene

An explosion in an enamelled agitator vessel was followed by a fire in a cooler of broken glass which was mounted on top of the vessel. The unit had been used for distilling water off toluene (boiling point 111°C) and during the several hours of cooling the agitator had been kept running. The temperature of the vessel at the time of the accident had fallen to 19°C. This is important because at this temperature toluene vapour and air form the most easily ignitable mixture (see 1.3.3). In addition to the now water-free toluene, the vessel contained an ion exchanger resin (polystyrene) in the form of small beads. As it is well known that substances mixed with chargeable liquids, such as toluene, often give rise to high electrostatic potentials it seemed obvious that static electricity was the cause of the accident.

Because of the general concern over possible electrostatic

hazards arising during distillations, an experiment was done on a laboratory scale to find out if the stirring of the water-free toluene and, separately, of the water-free toluene with an added solid component could cause a hazard. To simulate the conditions of the enamelled vessel a glass agitator was used in a beaker flask, the latter having been coated with a grounded conductor on its outside surface. The charge was monitored by means of a small independently air-vented induction probe. The results were remarkable. On stirring the water-free toluene alone the charge produced was insignificant and barely detectable (20 V) by the probe. When a quantity of resin, corresponding to 0.2 per cent of the weight of the toluene, was added to the toluene the potential rose to 400 V on stirring the mixture. On increasing the amount of resin to 4 per cent of the weight of the toluene a potential of 1.8 kV was reached when the mixture was stirred. This experiment, which was reproducible, showed convincingly that stirring a mixture of resin and toluene can give rise to dangerously high potentials (>1 kV). It follows that in the real situation an electrostatic ignition would be quite within the bounds of probability. However, the question remaining was that of the manner and location of the incendiary gas discharge. In view of the probable high potential at the surface of the stirred liquid it should be expected that brush discharges between the liquid and any of the metal fixtures would have been possible. Such discharges may well have been capable of igniting optimally combustible hydrocarbon/air mixtures such as toluene vapour and air at 19°C.

*Suggested safety measures*

- Provided it is acceptable, on the grounds of purity, an antistatic agent should be used which is capable of increasing the volume conductivity of the mixture to at least 10 nS/m.
- Reduce the concentration of oxygen in the vessel to below 10 per cent by venting with an inert gas.

## 5.9 Pumping polluted toluene

Minor faults may sometimes cause devastating results.

A new plant was built with the economical and ecological purposes of recycling polluted toluene. The plant had to be started before the pipeline to the liquid recycling station was finally installed. This meant that the polluted toluene, which was stored in a tank on the ground floor, had to be disposed of by pumping it into metal drums. Lorries then transported the full drums to the recycling station and returned them empty. A worker was responsible for preventing the collecting tank from becoming full because any overflow would automatically shut down the entire process.

A pump was installed to convey the toluene into a steel drum via a conducting hosepipe. Electrostatic hazards should have been avoided as the worker was wearing conducting shoes and was grounded by his standing on a steel grating floor. The drum also rested on the steel grating. By this means the possibility of spark discharges from the worker or the drum were eliminated (Fig. 5.4).

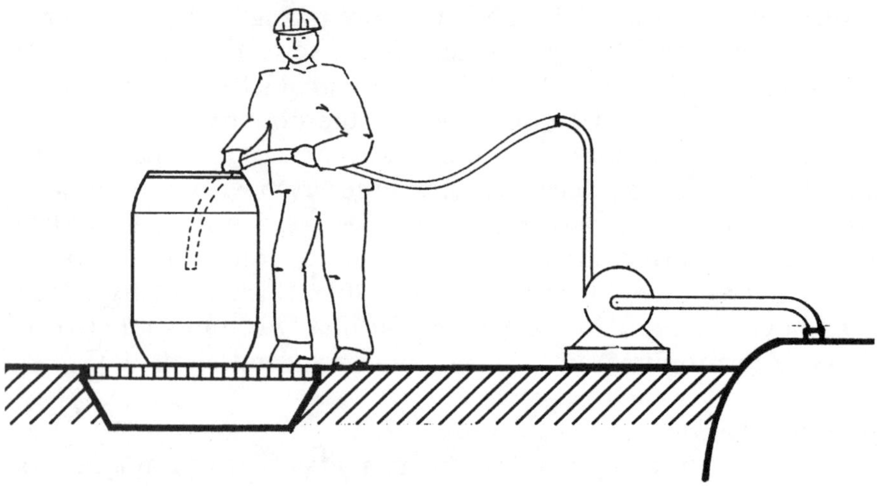

**Figure 5.4** Toluene being pumped into a plastic drum

During a nightshift one of the lorries returning the empty drums was delayed and this started a chain of events which finally led to disaster. The worker, having filled the last available metal drum, was faced with the situation that the overflow mechanism would operate when the tank became full resulting in a shut down. Wanting to avoid this he looked for other empty drums in the vicinity and found one made of polyethylene. He used it without being aware or caring that insulating materials can present an electrostatic hazard in that situation. The hosepipe was hung inside the plastic drum and the pump switched on. Shortly afterwards a flame shot out of the drum. The worker immediately ran for a water hosepipe and tried to extinguish the fire by pouring water into the drum. Unfortunately, this did not put out the fire but instead displaced the burning liquid over the rim of the drum causing a spread of the flames. He then gave up his own fire-fighting efforts and ran off to call the fire brigade without switching off the pump! As a result of the steady flow of toluene the fire spread to the nearby full drums which also caught fire. On arrival the fire brigade could not extinguish the blaze so they concentrated on protecting the surrounding buildings. The new plant was completely destroyed by fire! No further investigations seemed to be necessary. However, it is well known that when toluene flows through a pipe it becomes electrostatically charged. When the liquid is fed into a grounded metal drum the charge on it would readily be dissipated to ground (the volume resistivity of this polluted toluene was $1\,G\Omega\,m$). When feeding the liquid into the plastic drum the charge would build up to such an extent that brush discharges could occur between the conducting hosepipe and the surface of the approaching liquid which could ignite the toluene/air mixture. This was later shown to be the case by experiments.

*Suggested safety measures*

- Either warn the staff not to use plastic drums with combustible liquids, or, better, keep plastic drums out of the plant.
- Above all, people must be taught how to extinguish fires. Never use water with burning liquids which are not miscible with water.
- In this case the fire could have been smothered just by putting the lid on the burning drum.

# Chapter 6
# Case histories related to cone or bulking discharges

## 6.1 Plastic foam released from an autoclave

A spontaneous ignition occurred when a foam consisting of a mixture of isobutylene and a polymer was emptied, under pressure, out of an experimental autoclave into a steel container placed underneath. As no ignition source could be identified an electrostatic discharge was thought to be the possible cause of the accident. When the usual checks of the resistances to ground of the autoclave, container, floor, and the safety shoes worn by the workers had been made the results were always below 100 MΩ, thereby ruling out any possibility of a spark discharge. However, as the volume resistivity of the mixture was about 100 GΩ m its susceptibility to static charging was to be expected. Yet, there seemed to be no plausible explanation for an incendiary gas discharge.

We decided to do another test run but this time using pressurized nitrogen for the supply, to avoid an ignition of the mixture. The electrical field strength at the surface of the emerging stream of foam was shown to be −100 kV/m but no discharges were observed. In a further run at a higher nitrogen supply pressure the field strength rose to −150 kV/m and was accompanied by visible and audible radial gas discharges across the surface of the foam which had settled in the container. As was expected, there were no ignitions in the presence of the nitrogen gas and the experiment was terminated.

It was clear that the gas discharges being observed were not of the brush type nor could they be propagating brush discharges because the development of a double layer of charge in that situation was not possible. As cone discharges were unknown at the time, we were in the position of seeing incendiary gas discharges of a type we could not identify and with no idea of how to avoid them. Our safety measures, therefore, could only deal with the symptoms.

Our present view on the matter is that this was one of the first observed cone or bulking discharges related to liquids.

*Suggested safety measures*

- Ignitions can be prevented by using nitrogen gas pressure for the supply and for purging the container.
- The cone discharges could be prevented by the use of conducting and grounded dip rods.

## 6.2 Dust explosion in a silo

With regard to safety procedures in relation to the ignition of extremely sensitive combustible dusts, a new philosophy has come about in the last few years. Whereas previously it was generally considered that the exclusion of ignition sources was the only reliable means of preventing explosions, now there is gathering acceptance of the view that, apart from this, it is necessary to minimize the dangerous effects of an ignition. For this purpose, shock pressure resistant types of construction and the means of automatic explosion suppression are used. However, there is no agreement yet as to what values of MIE (1–10 mJ) for dusts are to be classified as extremely sensitive. These considerations were taken into account when a dust silo in a new plant was being planned, because the MIE of the product was known to be in the range 1–3 mJ. The safety measures adopted for the silo were to use a construction which was shock pressure resistant combined with pressure relieving panels on top of the silo.

After the plant had been running for about a year an explosion actually occurred in the silo when it had been pneumatically filled with organic powder to about 60 per cent of its capacity. Because of the safety precautions, no damage was done to the silo but the product was spoiled by the fire. This was reason enough to require an investigation into the cause of the accident. The silo, which was 3 m in diameter and 8 $m^3$ in volume, was made of stainless steel and was grounded. Of course, an electrostatic discharge was first suspected as being the cause of the accident because of the high volume resistivity ($>1\,T\Omega\,m$) of the product and, consequently, its ability to retain static charges. However, an investigation showed that all the conducting parts of the plant, including the filters, were reliably grounded thereby excluding any kind of spark discharge. Propagating brush discharges were also excluded because there were no insulating materials such as plastic coatings and tubes in the system. As brush discharges are not capable of igniting dusts (see 3.4.1) our attention was directed to cone discharges which are still a matter for research.

Before restarting the plant and to furnish proof of the existence of cone discharges we installed a number of capacitors close to the inner wall of the silo which would store any charge passing from the powder heap to the wall. Pulses of charge were, indeed, received by these capacitors and were monitored by means of computerized equipment for several months. The results showed that a single pulse of charge with sufficient energy to ignite a sensitive dust/air mixture happened about every fortnight. Even so, no ignitions occurred, implying that the cone discharges were not reaching places where optimally ignitable dust/air mixtures were present.

This case is a personal communication from Dr M. Glor who will be publishing the results in Poitiers (France) at the next European Conference on Electrostatics in 1997. The appropriate safety measures for this case will be discussed at the conference.

# Chapter 7
# Case histories related to propagating brush discharges

## 7.1 Plastic tube used in the pneumatic conveying of powder

In a spacious plant used for grinding synthetic organic material an explosion occurred which had devastating results. Because the powder was a combustible material, precautions against its possible ignition had been taken into account when the plant was built. The impact mill was protected by purging with nitrogen and was isolated from the rest of the system by a rotary valve. The temperature of the bearings of the drive shaft was constantly monitored to avoid overheating. Lastly, all conducting parts of the plant were reliably grounded to prevent the discharge of electrostatic sparks from them.

The events leading up to the explosion were as follows. After a period of use a stainless steel feeding pipe which led to the sieving machine became blocked through overloading, resulting in a time consuming standstill of the plant. In order to be able to see in time when a blockage was starting to build up, a 1m length of translucent plastic tube was inserted into the stainless steel pipeline. After checking the grounding of the metal pipes the plant was started up again. It was now possible to check the flow of the powder and make appropriate adjustments to stop any blockages.

Shortly after, workers in the vicinity of the plastic tube complained of being startled by electrical shocks. This was, no doubt, caused by the electric field from the charge on the inside of the tube produced by the moving powder (see 2.4), inducing charge onto the nearby workers. Because of this problem someone had the idea of wrapping a grounded copper wire helically around the plastic tube, thinking that it would dissipate the unwanted charge without obscuring the view of the inside of the pipe. The idea worked immediately and the person who thought of it earned the gratitude of his fellow workers until, unfortunately, a few hours later there was a loud explosion followed by several others in the branched system. The result was not only that the whole plant had to be shut down but that some of the workers required to be taken to hospital! To the astonishment of the plant manager, the electrostatics expert, without hesitation, diagnosed the problem as one of propagating brush discharges within the plastic tube (see 3.3.5). The effect of the copper wire was to direct the field from the charge on the inside of the tube outwards to the wire, thereby allowing a very high charge density to build up inside the tube until the breakdown potential of the wall of the tube was reached. At this stage a discharge would occur through the tube to the wire which would result in a propagating brush discharge inside the tube.

However, in the light of more recent knowledge, there is an argument for the case that the accident would have occurred eventually whether or not the copper wire was fitted to the tube. The electric field coming from the charge in the tube would attract counter charges in the form of ions and charged dust to the outside wall of the tube thus creating a bipolar system of charges. This could lead to the formation of very high charge densities on both the inside and outside walls of the tube and the eventual release of a propagating brush discharge. For this to happen inside the tube would not require an electrical breakdown of the tube itself, but merely a discharge inside the tube at the junction with the grounded metal pipe where the field strength would be very high. The only way to avoid propagating brush discharges inside the plastic tube would be to insert into it a grounded metal wire. Unfortunately, because of constant abrasion, the wire would be worn away in time and would become unreliable.

*Suggested safety measures*

- Insulating plastic tubes are not permitted for use in the pneumatic transportation of combustible dusts.

## 7.2 Plastic pipe used in the pneumatic conveying of powder

In a pneumatic conveying system used for transporting acrylic powder from a silo to a railroad bulk container the last few metres of a 50 mm diameter steel pipe had been replaced by a PE pipe for easier handling. On a rainy and snowy winter's day, after running the conveyor for only a few hours, an explosion occurred in the container. The only possible ignitable mixture in the container was that of acrylic powder and air.

On starting our investigations we found that every metal part of the system (pipe, valves, bulk container, etc.) had been grounded except a 150 mm long metal coupling between two sections of the PE pipe. The resistance and capacitance to ground of the coupling were found to be greater than 10 TΩ and about 12 pF, respectively. It is well known that powders can become electrostatically charged during pneumatic conveying and, because of this, we focused our attention on the charging of the isolated coupling, assuming that the latter had been the source of the incendiary spark. We estimated a potential of 10 kV on the coupling, corresponding to a stored charge energy of 0.6 mJ. Of course, this latter value appeared to be rather small for the ignition of a dust/air mixture but, as at the time there was no other conceivable explanation, the coupling was accepted as the source of the ignition. The solution to the problem was, therefore, to ensure that the coupling was reliably grounded.

Having repaired the pipe, grounded the offending coupling and replaced the damaged rail car, the conveying system was started up again. To our horror we received an urgent phone call informing us that after running it for several minutes there had been another explosion! We were astonished and perplexed at the news and knew that we had to act quickly if we were to make the system safe and maintain our credibility.

We began by running the system again but with nitrogen

blowing through it instead of air. By monitoring the field strength at several places on the rig and detecting any gas discharges with a radio-frequency receiver we hoped to locate the source of the ignition. In the event it turned out that these tests were unnecessary as we could actually see flashes inside the translucent PE pipe where it was connected to the metal pipe of the rail car. The flashes were almost a metre in length and were accompanied by audible 'cracks'. These discharges were what are now called 'propagating brush discharges', a highly energetic type of discharge, which we were not familiar with at the time.

The presence of snow and rainwater on the outside of the PE pipe provided it with an electrically conducting surface which was unintentionally grounded at various places. The charging of the inside surface of the pipe by the powder caused the outer surface to acquire a similar charge of opposite polarity, by induction, thus forming a charged double layer. In such cases each layer can reach a very high charge density. When the density of the charge inside the pipe was sufficient to cause an electrical breakdown, a discharge occurred over a large area of the pipe resulting in the release of a considerable amount of energy which was easily capable of igniting the dust/air mixture.

*Suggested safety measures*

- In pneumatic systems for conveying combustible powders the use of any electrically insulating material, particularly for pipes or the linings of metal pipes, is not permitted.

## 7.3 Plastic injector in a jet mill

A pinned disc mill was being used for crushing a coarse plastic powder and as the product had a relatively low MIE (5–10 mJ) the entire plant was, for safety, constructed to withstand high pressures. By this means neither the personnel nor the plant itself would be in danger. However, within a year of the plant starting up, an ignition occurred which, as expected, caused no injury and left the plant intact. The interior was found to be burned and encrusted with smouldering product the fumes from which were poisonous and therefore a danger to the maintenance crew. As the cleaning up of the plant was very costly and time consuming the accident proved to be a big loss for the company. Yet, it was only after two more explosions had occurred during the following year that the management and those responsible for health care decided to find out how to prevent further ignitions. Because traces of mechanical stressing were found in the pinned disc mill it was generally agreed that the crusher was the source of the ignition. It was, therefore, decided to replace the pinned disc mill with a jet mill as the latter had been shown to be free from ignition problems (see 8.10). After making this change everyone was amazed when, within two months, another explosion occurred. It, therefore, came as no surprise when with no other possible source in mind static electricity was blamed as the cause of the ignitions.

Investigations showed that the entire system was constructed from stainless steel and that every part was reliably grounded. However, it had been found that the plastic powder tended to cling to metal surfaces, thereby causing the blockage of the jet mill injector during the test runs. To prevent this and allow the injector to pass the coarse plastic powder into the mill, the injector had been lined with a polytetrafluoroethylene (PTFE) tube. This immediately directed our attention to the possibility of propagating brush discharges. In order to convince the staff of this we were permitted to install a small pressure-proof window into the feed system through which we could see inside the injector. What we saw surpassed all imagination! Sparks were crossing the entire length of the PTFE liner in rapid succession. There was no doubt that they had the potential to ignite the coarse powder which contained a quantity of fines. Being familiar with the mechanism of propagating brush discharges, remedial measures were easily achieved by experiment. Small holes of about 1 mm diameter were drilled in the tube wall at a distance from each other of about 20 mm, thereby limiting the area of continuously charged surface to less than about 500 $mm^2$. Initially, concerns that the effectiveness of the holes would be diminished because of encrustation were not confirmed, presumably because the dielectric strength of the encrustations was always much less than that of the undamaged PTFE. Propagating brush discharges did not occur again.

*Suggested safety measures*

- Linings of dielectric materials which can give rise to propagating brush discharges should be perforated at suitable intervals.

## 7.4 PE liner in a metal drum

A vulcanizing additive of caoutchouc, after being ground to a powder in a pinned disc mill, was passed through a pipe into a metal drum containing a PE liner. On being told that an explosion had occurred in the drum we checked various electrostatic properties of the system. The results were as follows:

- MIE of the powder: 5–10 mJ
- Volume resistivity of the powder: 100 TΩ m
- Field strength at the surface of the powder in the filled drum: 0.8–1.0 MV/m
- Potential to ground on an inserted probe of 20 pF capacitance: 25 kV
- Resistance between the drum and ground: $<1$ kΩ.

It follows from these results that the powder could sustain an enormous electrostatic charge, but that a spark discharge would not be possible because of the low resistance to ground of the drum.

We discussed the accident with Blythe (1) who informed us that charged powder collected in a container made of an insulating material will release its charge at the top of the powder heap due to the ionization of the air in that region. The effect of this is that ions of the same polarity as that on the powder are drawn to the inside walls of the container where they are collected, and ions of opposite polarity go to the powder, as is shown in Fig. 7.1. The field from the charge on the liner is directed towards the metal drum on which it induces counter charges. In this situation very high densities of charge can accumulate on the surface of the liner, and the drum, resulting in a strong polarization of the PE (dielectric) liner. Should the electric field inside the PE liner reach the breakdown value of the dielectric, a spontaneous propagating brush discharge could occur (see 3.3.5.1) which would, undoubtedly, cause the powder to ignite. However, for such an ignition to occur it is essential that the discharge through the PE liner is above the heap of powder, where there is a combustible powder/air mixture. The reason that ignitions seldom occur is thought to be because most discharges take place at the powder/liner interface. It is a fact that micropunctures in the PE liners are found after filling the drums when no ignitions have taken place.

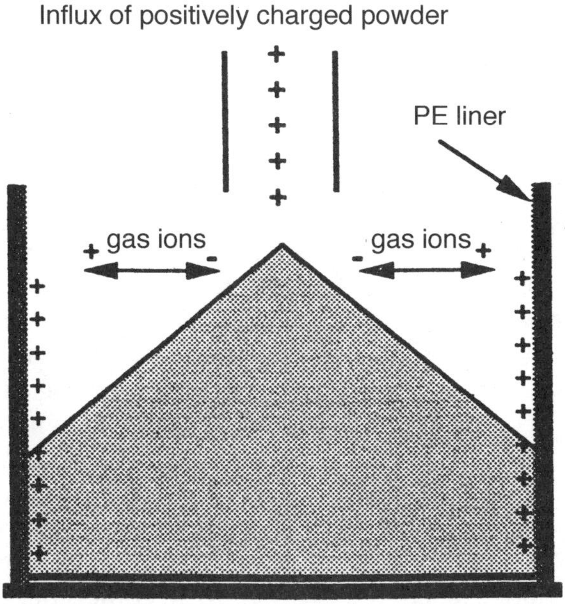


**Figure 7.1** Transfer of charge by gas ions from a heap of powder to the lining of a metal drum

*Suggested safety measures*

- During the filling of the drum the charge on the powder may be practically eliminated by mounting a grounded wire or rod at the collecting point of the powder. Although this method is proven, it was not adopted in this case because it is laborious to implement.
- As propagating brush discharges are reliably prevented by ensuring that the breakdown potential of the PE liner is less than 4 kV, the preferred safety measure is to prevent the potential on the liner reaching this value by the artificial microperforation of the PE.

Before investigating the case described above we had been called to two similar accidents concerning a pharmaceutical and a plastic product but failed to discover the sources of ignition. They each involved the fine grinding of the product and its transfer to a grounded metal drum containing a PE liner. In the light of the above case it would seem highly likely that the sources of ignition were propagating brush discharges.

## 7.5 PE liner in a paper drum

An explosion occurred when a pharmaceutical powder of MIE 5–10 mJ was emptied out of a paddle dryer (without rotating rods) into a cardboard drum containing a PE bag liner. The bag was fixed to the outlet of the dryer by means of a ring. The accident happened at the moment the worker was trying with his bare hands to close the top of the nearly filled bag which protruded above the drum. He was burned by a tongue of flame and the product caught fire. It was realized that the product was highly charged because of the sound of electrostatic discharges. Consequently, it was at first thought that the field from the charged powder had caused the worker to become charged by induction. This was later shown not to be possible as resistance measurements on the footwear and the flooring indicated a resistance to ground of not more than 100 MΩ.

At the time we were familiar with the case of the PE bag liner inside the metal drum (see 7.4) in which an accident was caused by a propagating brush discharge and wondered how we could do appropriate exploratory tests in the present case. It proved to be very difficult to initiate electrostatic discharges with the bare hands and at the same time identify the type of discharge. The unconventional idea of visibly trying to detect the discharges proved helpful. First, we purged the whole system with nitrogen to avoid any possible ignition. Then the area was darkened so that only the outlines of objects were visible to an adapted eye. What we observed was quite remarkable! As the powder fell into the bag iridescent light was observed across the surface of the powder. On bringing the hands close to the bag the same effect was seen at regions inside the bag near to the edges of the hands. When the hands were moved quickly towards the heap of powder, flashes of light were observed and shocks to the hands were felt. At the time we thought we were observing propagating brush discharges, but in the light of more recent knowledge they may well have been super brush discharges (see 3.3.3). In either case, because of the low MIE of the product, ignitions would always be possible.

*Suggested safety measures*

- Replace the air in the filling system with nitrogen.

## 7.6 Polymethylmethacrylate (PMMA) window pane in the silo of a granulating plant

In a polyethylene processing plant several production lines were installed and in order to even out any variations and maintain constant quality the entire product was fed into a mixing silo. As is shown in Fig. 7.2, the product was taken from the bottom of the silo and conveyed to the top for recycling. Because the product was clean PE granules no danger of fire or explosion from electrostatic discharges was expected and so no special precautions were taken. It was, therefore, almost unbelievable when an explosion occurred in the mixing silo which, fortunately, caused damage only to the silo. Even so, as was the routine, the electrostatics expert was called in to identify the source of the ignition and find out what had been ignited. An examination of the silo revealed that fines of PE, produced by the abrasion of the granules, had settled in the crevasse between the wall and the lid of the silo where it was sheltered from the mainstream flow of air. Such an accumulation of fines could, periodically, become dislodged and fall into the silo thereby causing a dispersion of ignitable dust. However, the question remained as to the nature of the igniting source and the coincidence of the discharge with the falling fines. Two more observations brought us closer to solving the problem. When the explosion occurred the silo was almost empty and a PMMA window in the silo was afterwards found to be partially covered with soot and its inner surface melted. The window, which was mounted in a conducting rubber gasket, was taken out and on inspection was found to have at least three tiny punctures which could only have been caused by propagating brush discharges (see 3.3.5.2). Part of the high charge density on the window necessary for a propagating brush discharge was doubtlessly caused by the continual impact of the PE

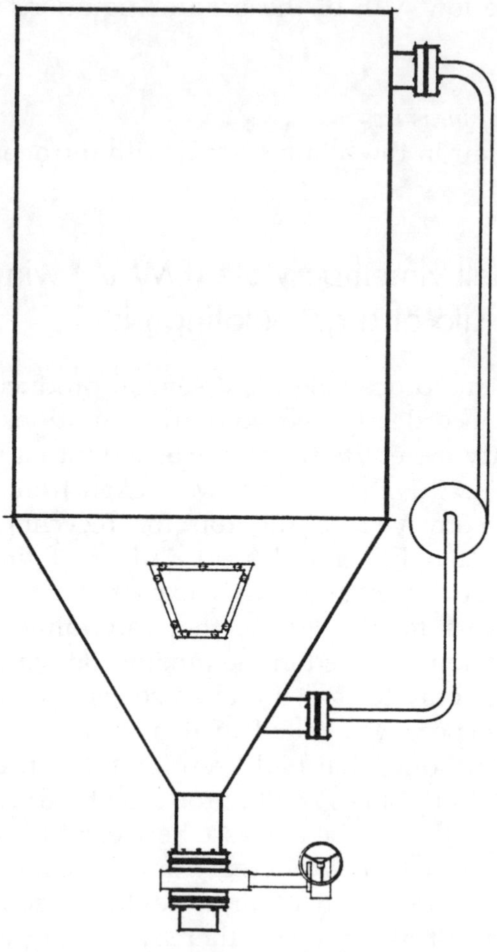

**Figure 7.2** Silo with a window pane of PMMA

granules with the window. The critical charge density was reached during the impact of the fines with the window, resulting in a propagating brush discharge and the ignition of the fines.

*Suggested safety measures*

- Even in systems where only clean granules are being transported, the presence of fines should be anticipated. The design of the silo should be such as to avoid regions in which fines can become lodged.

- Propagating brush discharges cannot occur with insulators of thickness greater than 10 mm (see 3.3.5.1).

## 7.7 PP coated expansion pipe

A suspension of solid matter in water, contained in a collecting vessel, was being forced pneumatically through a pressure strainer. On completion of the straining the excess pressure of ~3 bar (~0.3 MPa) in the vessel was released via a steel waste pipe with a polypropylene (PP) coating on the inside. At that instant an acoustic shock was heard and the pipe was blown apart, much to the consternation of the staff! A gas analysis of the suspension indicated the presence of ethylene oxide, which had been left over from an earlier polymerization process and was now in the pressurized air at a concentration just above the lower limit of flammability. The idea of an electrostatic discharge as being the ignition source was, at first, quickly dismissed because the pressurized air was very humid and therefore likely to militate against any electrostatic charging. However, laboratory experiments showed that water droplets running through an inclined PP tube acquired a high positive charge. The reason for this is that as the PP surface is not wettable (hydrophobic) charge separation readily takes place at the water/PP interface. During the release of the gas pressure very high flow rates were to be expected which could lead to high densities of charge on the PP coating. Brush discharges within the expansion pipe itself were not expected because there were no electrodes there, but such discharges may well have occurred at the junction of the collecting vessel and the pipe. Further, it was quite within the bounds of probability that during short periods of time double layers of high density charge were produced resulting in propagating brush discharges. These could easily ignite ethylene oxide/air mixtures which are close to the lower limit of flammability.

*Suggested safety measures*

- Plastic coated metal pipes should be avoided in the presence of combustible gas/air or vapour/air mixtures when airborne particles (aerosols), including water droplets, may be entrained.

## 7.8 Plastic tank inside a metal mould

Large plastic vessels with complicated shapes, e.g. petrol tanks for cars, are manufactured by tumbling molten plastic inside a metal mould. The mould comprises two parts which are hinged together and which can be rotated about two axes (vertical and horizontal) simultaneously. A plastic monomer is poured into the mould and while the latter rotates the monomer becomes polymerized at the heated walls of the mould, thereby forming the plastic container. The mould can then be swung open and the finished container removed.

After several trial runs in which the rotating speeds of the mould and its temperature were optimized, an explosion occurred inside the mould on the following day, as the system was still cooling down. The locking mechanism was blown apart and one-half of the mould struck a worker causing fatal injuries. It was already known that an explosive atmosphere would be present inside the mould because of the flammable vapour coming off the liquid monomer, but there was no question of there being any source of ignition. However, following the terrible accident there was a vague suspicion that an electrostatic gas discharge might have been the cause. As nothing had been changed after the accident it was possible to make a thorough examination of the remains. The first important observation was that the explosion had occurred during the cooling of the mould, after the polymerization of the plastic. Thus, the plastic container was already finished at the time of the explosion, and any electrostatic effect from the liquid monomer could be discounted.

The question, then, was how could electrostatic charging take place within a closed system in which there were no parts moving against each other? On examining the two parts of the container, by the use of a magnifying glass, a tiny pinhole with a melted rim was found at a flat part of the container. This alarmed the electrostatics expert because only propagating brush discharges leave traces in the form of pinholes. After repairing the mould several experimental runs were done using nitrogen instead of air inside the mould. As before, each plastic container was scrutinized for pinholes until one was found. It was again noticed that the hole was located at a flat part of the container where the plastic was particularly thin and indented towards the inside of the mould. On powdering the area of the pinhole with xerographic toner a pattern of the discharge channels of a propagating brush discharge was revealed. We were still wondering how a sufficiently high quantity of charge to cause a propagating brush discharge could have accumulated inside the mould.

Tests for the effect of temperature on the surface resistivity of the plastic showed that a rise of 40°C caused a reduction of about an order of magnitude. On separating the plastic container from the metal mould during the time of cooling, the cooler and therefore the more resistive outer wall of the plastic container could become highly charged. This charge would induce a similar charge of opposite polarity on the hotter and more conducting inner wall of the container. Thus, a double layer of high density charge could be formed in this way leading to an electrical breakdown of the plastic and a propagating brush discharge.

The cooling of the outer wall of the container would cause it to contract relative to the inner wall resulting in the formation of mechanical stresses within the plastic. At the moment when the temperature difference between the inner and outer walls of the container reached a maximum, the internal stresses would be at their greatest resulting in an abrupt separation of the container from the mould. At this stage an electrostatic discharge, as described above, would be released.

*Suggested safety measures*

- As a general solution to the problem, purging with nitrogen would normally be recommended. However, in this case, because of the fatal accident, the method was discontinued.

## 7.9 Literature

(1) Blythe, A.R. and Reddish, W. (1979). Charges on powders and bulking effects. *IOP Conf. Ser.*, No. 48, 107–114.

# Chapter 8
# Case histories related to spark discharges

## 8.1 Dusts

## 8.2 Rotating beater dryer (1)

On emptying rotating beater dryers into grounded metal drums placed underneath, ignitions occurred every now and again which sometimes caused flames to come from the outlet orifices, thereby endangering nearby workers. As the dryers were pressure resistant they did not suffer any damage, but the product was spoiled. The management of the drying plant were unable to find the cause of the fire, but it was observed that whenever a fire occurred sensitive materials (MIE 5–10 mJ) were being dried, and that the ignitions happened only when the dryers were almost empty. Being concerned about the danger to workers and the loss of product, the management finally called in the experts on electrostatics.

Before describing the investigation, the dryer and its mode of operation will be explained first with the use of Fig. 8.1. It shows the cross-section of the dryer which is about 5 m long and 2 m in diameter. Along the central axis of the dryer is a rotating shaft to which are attached several shovels used for stirring the product. When the dryer is half full of wet product and the feed opening has been closed, steam is admitted to a heating jacket around the

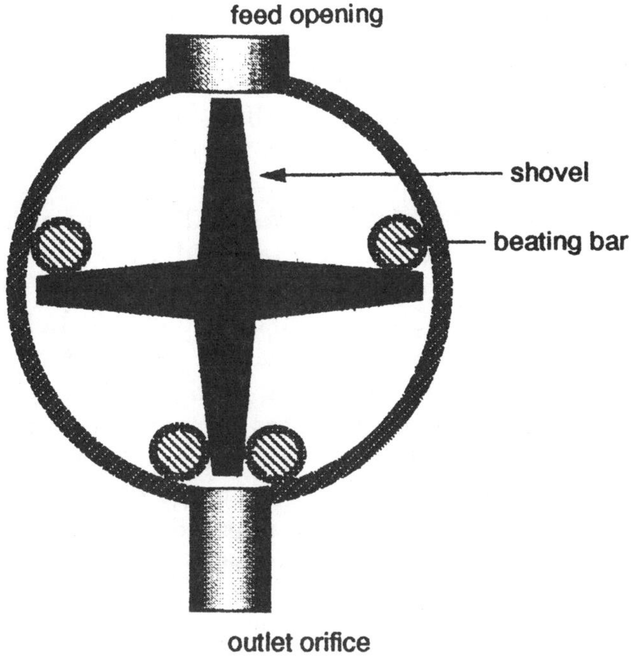


**Figure 8.1** Rotating beater dryer

dryer (not shown) and the moisture which is evaporated from the product is drawn off by means of a vacuum pump. When, after several hours, the product is dry the pump is switched off and the evacuated dryer is filled with nitrogen. The outlet orifice is then opened and the rotating shovels transport the product through the orifice into the drum. As the product leaves the dryer, air replaces it so that when the dryer is nearly empty there is a more or less combustible mixture of product and air in the dryer.

But what, in the case of this accident, was the source of the ignition?

As any kind of incendiary discharge emits radio-frequency signals (see 3.3.2), an AM radio receiver was installed in the region of the feed opening. As soon as the dryer was filled with nitrogen many radio signals were detected but still we did not know how they were caused. It should be noted that there are four steel beating bars of 100 mm diameter (Fig. 8.1) inside the dryer, which extend the entire length of the dryer and are moved by the rotating shovels. Their purpose is to break up encrustations and lumps of product. They may become coated with the almost dry but sticky product so that, depending on the resistivity of the product, the beating bars can become more or less electrically isolated from the grounded dryer. Values for the resistance between the bars and ground of up to 1.0 TΩ were measured, but often there was good electrical contact with the ground. The latter was, presumably, due to contact points between the shovels and the bars where the encrustations had been chipped off. However, there was good evidence to show that the bars do become charged as they roll around in the insulating product and potentials of up to 6 kV were measured on the bars immediately after stopping the shovels. The capacitance to ground of the bars was about 600 pF. By using the equation for the energy of a spark (see 3.3.6), it was estimated that the energy released in a single discharge would be about 10 mJ. Thus, the evidence strongly indicated that the source of the ignition was a spark from the beating bars. Of course, this could only cause an ignition when sufficient oxygen from the air was present, i.e. when the dryer had been emptied.

*Suggested safety measures*

- As it is impossible to reliably ground the beating bars (e.g. by cables), the only safe measure available is to purge the dryer with nitrogen as it is being emptied.

## 8.3 Explosion of a resin powder in a metal drum

In a small plant a multi-purpose powder mill was used for grinding various types of plastic material. The mill was situated on the upper floor of the plant and the ground plastic was fed through a grounded metal pipe to a metal drum in the basement. To avoid an electrostatic hazard the drum had to be reliably grounded when being filled. The idea of manually attaching a grounding clamp to each drum was rejected as it was thought the workers might occasionally forget to do it. Instead the plant engineer arranged for a grounded metal plate to be fixed to the floor on which the metal drums were rested.

This worked well for a long time until, one day, after about 20 drums had been filled with a crushed resin, an explosion occurred inside a drum which blew the lid off. What was the cause of the explosion?

As no other fires were detected in the mill at the time, it was thought that an electrostatic discharge was the likely cause. On inspecting the site of the accident it was noticed that the bottom rim of the drum had left an imprint on the grounded metal plate. This was caused by the hot drum melting resin which had been deposited on the plate before the explosion occurred. The worker confirmed that resin had accumulated on the plate while the drums were being filled. He also told us that he had had difficulty in removing spilled resin from the plate with the use of a vacuum cleaner.

After placing another drum on the metal plate we determined its resistance and capacitance to ground. The values were greater than 1.0 TΩ and 380 pF, respectively. As the volume resistivity of the resin was greater than 1.0 TΩ m, it became apparent that the spilled resin had formed an insulating layer on the surface of the plate. Tests on a drum as it was being filled with resin indicated that the latter was charged and that a potential of at least 8 kV was induced onto the drum before an electrical breakdown occurred between the drum and the metal plate. By substituting these data into the energy equation, the energy of the charge stored on the drum was found to be about 12 mJ. The time constant of the system was 380 s.

As the MIE of the resin was found to be between 5 and 10 mJ, it was clear that an ignition of the resin dust by an electrostatic discharge was quite possible. The question remaining was at which place could a spark have been passed through the dust?

The drums had metal filling lids with PU-foam gaskets on their undersides. When a lid was lowered onto a drum it formed a dust-proof seal. As the filling pipe and the lids were always reliably grounded, it seemed likely that a spark was passed from the upper rim of a charged drum to a grounded lid at a place where an explosive mixture of resin dust and air was present.

*Suggested safety measures*

- While a fixed grounding system is preferable to a hand operated one, it is important to check its efficiency regularly.

## 8.4 Dust removal from tablets

Although many tablets consist of combustible materials, they do not, in themselves, present an explosion danger because of their size. However, during their production in the pressing machine dust is deposited on the tablets, albeit in only small quantities. Consequently, we were very surprised to learn that an explosion had occurred in an apparatus used for removing dust from tablets. As the apparatus was very simple and without any electrical attachments it was thought from the start that static electricity was the cause of the accident. The task of the electrostatics expert was, therefore, to determine the cause of the ignition and to identify the material which had been ignited.

The cleaning apparatus consisted of two main parts, a transportation box with a perforated mezzanine and a silo-like construction with a tangential air outlet at the top. Each was made of metal. The tablets to be cleaned were poured onto the mezzanine and the box was placed beneath the silo. A ventilator attached to the outlet of the silo was switched on causing air to be drawn upwards through the tablets thereby carrying the dust into the silo and out at the top. Despite the small quantity of dust, tests on a similar dust remover had shown that the charge on the dust could readily cause an isolated box to become charged by induction to a potential of about 6 kV.

For the purpose of concentrating the flow of air through the tablets the contact between the box and the lower rim of the silo was made air-tight by means of a rubber gasket. As the gasket and the rubber wheels of the box were of electrically non-conducting materials, charge on the box could accumulate until a spark occurred between it and the grounded silo. Thus, a possible ignition source was identified but, because the concentration of dust was too low to be ignited, the problem of what had actually happened was still unsolved.

As the damaged apparatus had already been dismantled we examined a similar unit which was still running and found that near the top of the silo there was a region of little disturbance of the air where a lot of dust had settled. In our view it was likely that a quantity of this dust had become dislodged in the damaged apparatus and had been swirled up at a place between the transportation box and the rim of the silo where a spark discharge had occurred.

*Suggested safety measures*

- The formation of combustible dust atmospheres is sometimes difficult to predict but the possibility of such occurrences cannot be ignored.
- Ensure that the transportation box, as well as the silo, are each reliably grounded.

## 8.5 Filter bag with a supporting framework

When drying dusts, e.g. in a conveyer flash-dryer, it is always necessary to separate the dust particles from the drying and conveying air at the end of the process. This may be done by using, for example, a cyclone. However, these fine dusts which are always present have to be filtered out by using filter bags of fabric which need to be cleaned regularly to maintain their efficiency. Among the different ways of cleaning filter bags *in situ* the most effective is by using pulses of pressurized air. This always requires some form of support for the bags. In the present case metal baskets were used because of their high mechanical strength.

A new conveyer flash-dryer plant equipped with a filter bag system which could be cleaned *in situ* by pressurized air pulses was set up. Although everything had been carefully planned, it was not long before an explosion occurred which destroyed the filter. At the time it was assumed that the ignition source could only have come from inside the filter housing. But just think about it! There were no moving parts inside the filter, indeed the entire system appeared to be fixed. So how could the accident have been caused by static electricity? After examining the system we concluded that a spark discharge might well have been the source of ignition. By referring to Fig. 8.2 it can be seen that the filter bag with its supporting basket (about 2.5 m long) is fixed to the filter holder in the carrier plate. To achieve reliable filtration the filter bag is folded down inside the basket. The pulse of pressurized air is applied to the filter via the concentric cleaning nozzle thereby shaking the dust off the fabric.

When checking a filter bag of similar construction it was found that the metal framework of the filter was insulated from earth by

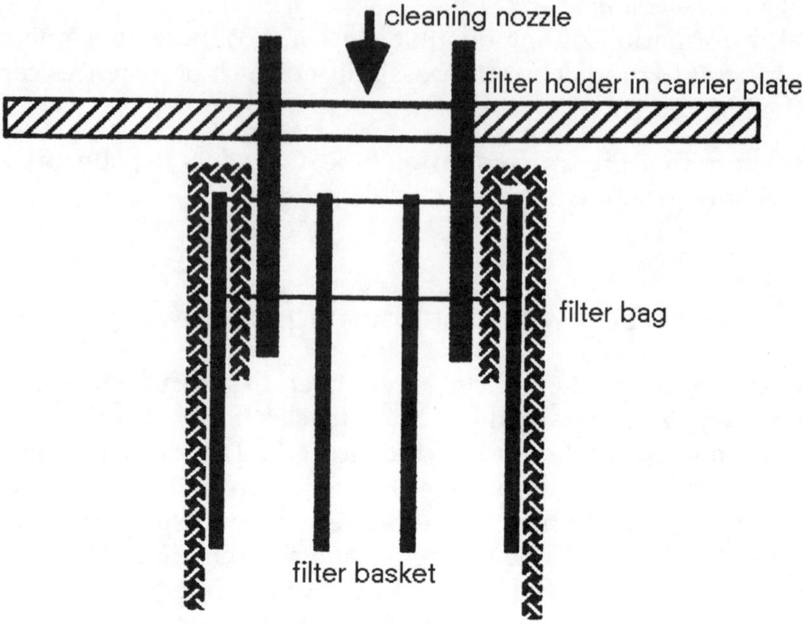


**Figure 8.2** Filter bag on a supporting framework

the bag. Electrical tests on the metal basket yielded the following results: capacitance to ground, 200 pF; resistance to ground, 1.0 TΩ. A charge was built up on the filter bag as the dust settled on it. The electric field from the charge on the bag produced a similar charge and a corresponding potential on the isolated metal basket, by induction. In this case the voltage was found to be insufficient to cause a discharge to nearby grounded conductors. However, at the instant the pulse of pressurized air was initiated a strong spark occurred between the metal basket and the carrier plate causing a perforation of the filter bag. It was estimated from the length of the spark that the potential on the basket was in the region of 15–20 kV. By applying the energy equation the corresponding energy was, at least, 20 mJ and was sufficient to ignite the dust.

*Suggested safety measures*

- Ensure that the metal basket is reliably grounded. This may easily be done by sewing a conducting ribbon to both sides of the open end of the filter bag so that when the fabric is folded down there is good electrical contact between the filter basket and the grounded filter holder in the carrier plate.

## 8.6 Filter fabric containing interwoven steel fibres

An explosion was reported in a dust filter system of the type in which separate filter bags (hoses) were suspended from the ceiling of the caisson and the dust-laden air collected on their inner surfaces. The dust which exploded was a special mixture containing a large amount of caoutchouc which had rarely been produced up to then. After the rubble had been examined and cleared away, the management made it clear that there had been no ungrounded metal parts in the filter. The experts were completely puzzled and had no idea as to how the ignition could have occurred. Several days later, while the experts were still thinking about the accident, the workers were shocked by the news of a much stronger explosion in a similar filter system on a nearby silo. The dust which was being filtered at the time was, again, one containing a substantial amount of caoutchouc.

During later discussions it was mentioned that some years ago filters of fabrics containing steel fibres had been installed in the plant to minimize the danger of static electricity. However, with the passage of time it was felt that this measure had not proved to be effective, presumably for reasons which were not understood, and that as the bags deteriorated they were replaced with ones of fabric without steel fibres. Even so, it was not certain that some of the bags still in use were of the type with steel fibres.

Previous tests on the bags with steel fibres had shown that their surface electrical resistances increased from about 2 kΩ to 70 TΩ as the distance between the electrodes was increased from a few millimetres up to 300 mm. These results showed that there was no overall electrical connection between the steel fibres, indicating that there could be no reliable dissipation of charge to ground.

In the light of these findings we decided to examine in the laboratory the conditions, if any, in which electrostatically charged filter bags could cause the ignition of the dust. Two filter bags were used, one with and one without steel fibres in the fabric.

Both were washed to remove any conducting contaminants, then rinsed in desalinated water and dried. Metal clasps were attached to the upper and lower ends of the filters and the lower ones were grounded, as was done in practice. The upper clasps, which were isolated from ground, were connected to an electrostatic voltmeter during the tests. The bags were charged by means of a corona discharge from a pointed electrode connected to a high voltage source (+/−15 kV) via a resistance of 1 GΩ, and mounted half way up the bag at a distance of 150 mm from it. The results with the bag not containing steel fibres showed a slow increase in the potential of the upper clasp to a maximum value of 3 kV after one minute. This potential was the same when the bag was charged positively or negatively. On bringing a grounded metal ball up to the surface of the charged bags no discharges were observed. However, in practice, when highly charged dust is collected by the bag it is likely that much higher potentials than 3 kV would be achieved and that static discharges might be possible.

In the case of the filter bags with steel fibres the purpose of the fibres is to conduct charge from the bag to ground via the grounded clasp. But this is not possible when, as with the bag under test, the fibres are not electrically connected overall. On applying the corona discharge to the bag with steel fibres the potential of the upper clasp immediately rose to about 10 kV when the charge was positive. On attempting to charge the bag negatively, corona discharges from the steel fibres occurred immediately and the potential of the upper clasp did not reach 3 kV.

When the grounded metal ball was brought up to the positively charged bag a small spark discharge occurred and several steel fibres were seen to become red hot.

In another test with the bag containing steel fibres a small amount of combustible powder was spread on the surface of the bag. After charging the bag positively the metal ball was brought up to the powdered surface whereupon a flame was produced spontaneously. This test was repeated several times causing a number of ignitions, but not on every occasion.

Similar tests with bags not containing steel fibres were unsuccessful in causing ignitions.

The above tests demonstrated that filter fabrics with a heterogeneous distribution of steel fibres can cause weak spark discharges which, in themselves, are unlikely to cause the ignition of dust/

air mixtures. However, the type of steel fibre used in the bags became red hot during sparking resulting in the ignition of dust on the fabric. Although there was no evidence remaining after the explosions discussed earlier, because the filters were destroyed by fire, the heating of the steel fibres may well have been the cause of the accident.

Unexpectedly, the results of further tests to determine the charging behaviour of powders which were conveyed pneumatically into various silos strongly supported the hypothesis of an electrostatic ignition. It turned out that each of the powders which had been processed previously became negatively charged. But, as had already been shown, negatively charged filter fabrics containing steel fibres do not rise to a high potential because of the effect of corona discharging. However, it was found that the special powders which had been involved in the explosions and which contained a high amount of caoutchouc were positively charged. Thus it became clear that these powders could have caused dangerous levels of charge to build up on the filters leading to electrostatic ignitions by the mechanism which has already been described.

*Suggested safety measures*

- The use of steel fibres in filter bags is advisable only provided they are electrically connected throughout the material and can be reliably grounded.

## 8.7 Explosion when emptying a metal drum

It was at the end of winter and the first shift had started work at the plant after the weekend. When a steel drum filled with an organic powder was emptied into a mixing unit an explosion occurred at its opening. The drum was held by a fork-lift truck with a special device for holding and turning the drums (Fig. 8.3). Eyewitnesses reported that the explosion happened when the drum was almost empty and turned at such an angle as to touch the opening of the mixing unit. The first drum of powder to be mixed had been emptied into the clean and dry mixing chamber and it was considered that the only likely sources of ignition were an electrostatic spark or a mechanical spark caused by friction.

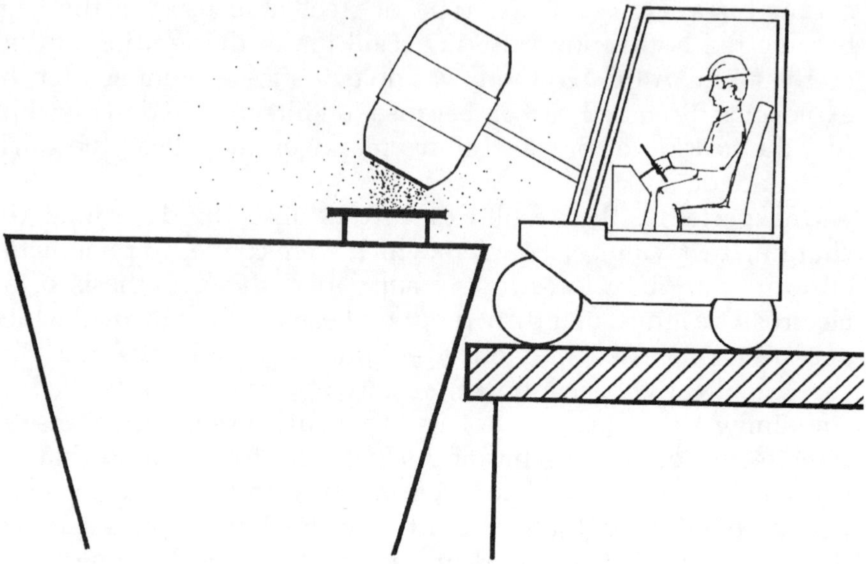

**Figure 8.3** Organic powder being emptied into a mixing unit

We looked first at the mixing mechanism. All the parts were made of stainless steel and were intact, and no foreign bodies were found. The entire mechanism was in excellent working condition and was free of damage and scratches. This left static discharges as the only suspect, although the work's manager was of the opinion that every precaution against possible static discharges had been taken. The truck was equipped with conducting rear tyres and the drum holder device had conducting rubber surfaces. The ceramic tiled floor had been tested years ago when the plant was new and had been found to be electrostatically dissipative, i.e. with a resistance through to ground of less than 100 MΩ. So it was to be expected that the drum would always be grounded. Experts in electrostatics tend to be sceptical of claims that ceramic tile floors are always electrostatically dissipative. On the contrary, ceramics can be excellent insulators, e.g. when used on overhead power cables. Conducting ceramics are an exception and their conducting behaviour depends on the amount of water absorbed by their pores. Experience with ceramic tile floors in the chemical industry indicates that their electrostatic dissipativity relies on the fact that they are often cleaned with water. However, for environmental reasons and in order to save water, floors are preferably dry-cleaned nowadays. This happened in the present case and as the plant remained heated over the weekend the floor could have dried out even more than usual. Tests for resistance to ground through the tiles gave values of up to 1.0 TΩ. The truck holding the drum in the emptying position showed a resistance to ground of 100 GΩ and a capacitance of 600 pF.

Tests in which a drum was emptied in about 5 s (as in practice) showed potentials of up to 13 kV on the drum/truck system for short periods of time. According to the energy equation, a spark discharge between the drum and the opening of the mixing unit could release energies of up to 50 mJ. This is about twice MIE of the dust.

*Suggested safety measures*

- It should not be assumed that a newly laid ceramic floor which is shown to be electrostatically dissipative will remain so thereafter.
- In general, it is recommended that all conducting equipment and isolated parts be grounded individually, rather than relying on conduction through the floor.

## 8.8 Filter fabric made partially conducting by a flame-proofing agent

Safety measures are sometimes taken which are beyond those required by the appropriate standards, as a means of ensuring 'belt and braces' protection. However, it is a fact that added precautions can sometimes lead to accidents. In the following case two extra precautions had been taken than were necessary, and there was great concern when it was discovered that it was a combination of these which brought about an accident. The following describes the course of events as they happened step by step.

A powder of low molecular weight polyethylene was pneumatically transported into a 5 $m^3$ silo on top of which was a filter housing. The filter itself was comprised of a number of filter bags each mounted on a metal basket. Sufficient safety precautions to avoid any spark discharges had been taken by reliably grounding all conducting parts of the filter. As the powder was of a low MIE (5–10 mJ) someone had the idea that a higher degree of safety could be achieved by using bags containing interwoven conducting fibres. The purpose was to prevent brush discharges coming from the fabrics, despite the fact that it is known that such discharges do not ignite dusts (see 3.4.1). The filter fabrics were tested according to DIN 54345 Part 5, i.e. the resistance was measured across a 300 mm length of fabric. As the resistances of the bags were always below 100 MΩ, they were installed into the filter housing. Grounding the bags was achieved by clamping them to the filter holders in the carrier plate. Before starting up, the resistance to ground of each filter was checked and found to be satisfactory, thereby putting the minds of the operators at rest.

Later another idea was adopted, namely, that of using filter bags which had been treated with a flame-resistant finish in order to prevent a fire should an ignition occur. A manufacturer was found who offered filter bags with a flame-resistant finish and which also contained steel fibres. These bags were used to replace the original ones and resistance tests on them showed that they were satisfactory. Now the operators, with their minds fully at rest, could sleep even better!

These changes were made in the early summer. On the first frosty night in November an explosion occurred in the silo which was met with astonishment and incredulity by the management and the workers! Fortunately, because of their flame-resistant finish, the filter bags were only partially destroyed and it was possible to measure their resistances across 300 mm lengths of material. To our surprise they were found to be greater than 10 TΩ and we were puzzled to know how this could be.

Unused bags of the same type as those used at the time of the accident were tested after being dipped in water and dried to equilibrium in an atmosphere of 10 per cent relative humidity. The same high resistances were found as before. Further investigations revealed that the flame-resistant finish was ammonium phosphate which imparted an acceptably low resistance to the fabrics under normal humidity conditions. The fact that the unused bags, after being rinsed in water, still had high resistances indicated a lack of electrical continuity between the steel fibres. This latter deficiency had, therefore, been masked by the effect of the finish on the conducting properties of the bags. Even so, these filter bags would only become electrostatically charged when they were fairly dry. Such was the condition of the bags in use. As the air used for the pneumatic transport of the powder became heated under compression and the relative humidity of the frosty air outside was already low, it follows that the heated air would be drier still. Only brush discharges would be released by the bag material itself and these would not be capable of igniting dusts. However, tests showed that the separated steel fibres formed isolated conductors of lengths up to 200 mm on which charge could accumulate. The ignition of the dust was attributed to spark discharges between the steel fibres and the grounded filter basket.

*Suggested safety measures*

- The electrical interconnection and grounding of conducting fibres in filter fabrics must be reliable.
- The required conducting properties of the fabrics should not depend on the use of fabric finishes.

## 8.9 Emptying a tumble dryer

On pouring a vulcanizing additive from a tumble dryer into a silo an explosion occurred when the dryer was almost empty. As the additive was a solvent-wetted product it was suspected that a brush discharge from the possibly charged product itself was the cause of the ignition. However, a thorough investigation showed that the product had been dried under a vacuum of 6 mbar (0.6 kPa) after purging the atmosphere with nitrogen. An explosible solvent vapour/air mixture could not, therefore, have been present. Further investigations revealed that the dry additive was an easily ignitable dust with a MIE of about 10 mJ. As there is no evidence to date of dusts being ignited by brush discharges (see 3.4.1) we had to look for other possible sources of ignition in the silo.

The tumble dryer was emptied via a grounded conducting pipe into the metal silo (Fig. 8.4). The lid was not fixed to the silo and could be easily moved. Also, there was nothing connecting the lid and the emptying pipe from the tumble dryer.

Following the explosion the resistance between the lid and ground was measured and found to be about 100 kΩ. This is

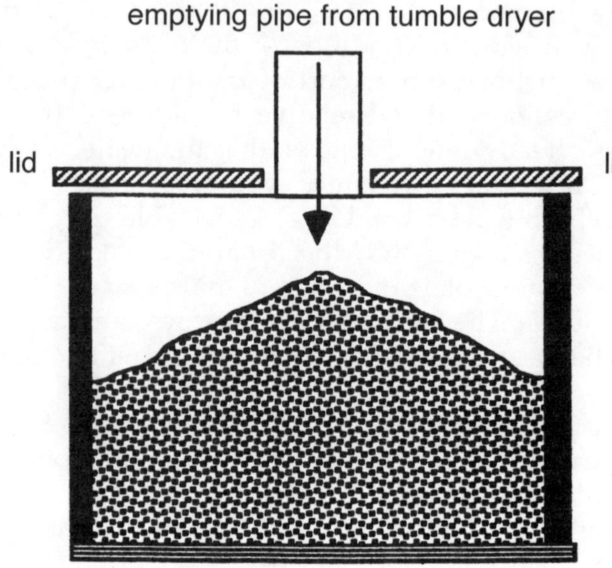


**Figure 8.4** Silo with a moveable lid

sufficiently low to prevent any possibility of spark discharges from the lid. However, on reconstructing the incident it was shown that the lid could have been isolated from ground by product deposited between the lid and the rim of the silo. When the emptying pipe was positioned concentrically in the opening of the lid the resistance between the lid and ground was found to be about 100 GΩ. Unfortunately, the capacitance of the lid in this isolated position was not measured but it was estimated that it could store enough electrical energy to produce an incendiary discharge in the presence of the sensitive product. The question still remained as to how the lid could have become electrostatically charged without having any contact with the charged product.

During the emptying of the dryer the level of the settled product in the silo rises thereby causing the electric field from the product to be drawn more and more towards the lid, causing the latter to become charged by induction. It is not known whether a spark discharge from the lid could have passed to the rim of the silo or to the emptying pipe, but it is certain that an ignitable dust/air mixture was present in both places.

*Suggested safety measures*

- All conducting parts of a plant, including those which are not expected to make contact with the charged product, must be reliably grounded.

## 8.10 Cyclone separator set up on a drum

It may be that a job is already finished before it is started!

A combustible rubber additive had to be ground to a fine powder. The task was to find out if this would be possible with the use of a jet mill which, it was assumed, would not give rise to static discharges capable of igniting any dusts. The test equipment was arranged as follows. The additive was collected from the grounded jet mill into a metal drum which was also grounded. The airborne particles in the exhaust air of the jet mill were to be separated in an attached cyclone with a dust filter on top. The cyclone itself was also arranged on a grounded metal drum which was lined with a polyethylene bag. The latter was turned down over the rim of the drum to catch the fine dust. In order to make the mounting easy the cyclone was connected to the jet mill by a piece of soft rubber pipe. The jet mill was started by opening the air valve. Shortly after pouring in the additive an explosion occurred. The cyclone was thrown off the drum and, luckily, nobody was hurt. The smell of burned rubber additive was still in the air as the operators realized that a jet mill was not as safe as had been presumed. However, jet mills had been tested several times before at different test houses and had been shown to be safe from ignition problems. Before questioning this statement it was decided to examine the cause of the ignition again. The slightly damaged cyclone was put on another metal drum lined with a new polyethylene bag. When the soft rubber pipe was attached again it became clear what had happened. The cyclone had been isolated from ground by the liner and had become charged by the contact and separation of the swirling additive inside it. The spark could have occurred between the cyclone and the inner side of the drum's rim, thereby igniting the dust as it entered the drum.

The measured electrical parameters of the cyclone were as follows: resistance to ground, 100 TΩ; capacitance to ground, 250 pF.

The energy in a spark discharge required to ignite the finely ground additive was about 10 mJ. By using the energy equation it can be shown that this would be reached when the potential on the cyclone was about 9 kV.

*Suggested safety measures*

- Ensure that all the metal parts which might become charged are reliably grounded.

## 8.11 Fire caused by an antistatic PE bag

On pouring an organic powder from a non-conducting plastic drum lined with an antistatic PE bag into a mixing machine, via a funnel, an explosive flame occurred at the funnel. Tests on the powder/air mixture gave a value for its MIE of less than 1.4 mJ, indicating that the product was easily ignitable.

In discussions with the workers we were told that rainwater, which had collected in a hollow in the lid of the drum, could have run into the drum when it was opened. The water would have wetted the outer surface of the PE bag, which was unopened, thereby enhancing its electrical conductivity. Even without wetting, the surface resistivity of the bags was 100 kΩ. Such a bag, when inside the insulating drum, was estimated to have a capacitance to ground of about 100 pF. As the powder was poured out of the bag the powder and the bag would become charged, that on the latter residing on its highly conducting outer surface. Tests for the potential on a wetted bag as it was being emptied showed values of more than 5 kV when only one-tenth of its contents had been transferred to the mixing machine. Fearing a possible ignition, the pouring of the powder was stopped immediately.

We also discovered that the PE bag could slide out of the drum and produce a spark on making contact with the grounded bar grating of the mixing machine. By using the energy equation it was shown that an energy of at least the minimum value required to ignite the powder/air mixture could be produced on the PE bag after emptying only a few kilograms of powder from it. From the available evidence it seemed most probable that the accident was caused by a spark discharge between the charged bag and the bar grating.

*Suggested safety measures*

- Use only liner bags of antistatic PE with drums which are electrically conducting and grounded.

## 8.12 Pouring powder into an agitator vessel

An agitator vessel of volume 6 $m^3$ (Fig. 8.5) was partly filled with ethyl acetate via a grounded stainless steel funnel inserted into the manhole at the top of the vessel. An organic powder was poured through the funnel into the vessel from a metal drum of volume 60 l, which did not contain a plastic liner, by a workman who was standing on a wooden platform. When the drum was almost empty an explosion occurred in the region of the funnel. The ethyl acetate (flash point −4°C) was at room temperature, and a combustible vapour/air mixture had formed around the

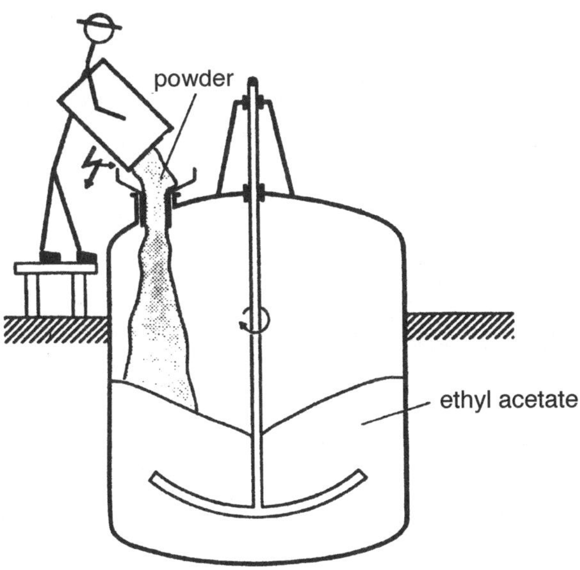


**Figure 8.5** Pouring powder into an agitator vessel containing ethyl acetate

funnel. The worker was wearing ordinary shoes of resistance 1 TΩ to ground and the resistance between the wooden platform and ground was found to be 200 MΩ. The capacitance of the worker to ground in his operating position was 160 pF. As, in this case, the resistance of the floor was negligible the relaxation time ($\tau = RC$) of charge on the worker was 160 s.

The powder, being a chargeable substance of volume resistivity greater than 10 GΩ m, became charged as it flowed out of the drum leaving behind charge of opposite polarity on the drum and the workman. Because of the high relaxation time of the man/drum system a spark discharge could have passed between the charged metal drum and the grounded funnel when they approached one another. From experience it was known that the potential of the drum could readily reach between 3 and 5 kV as it was emptied. By taking the lower value and using the energy equation, the energy stored on the system was shown to be 0.72 mJ. As this was greater than the MIE of ethyl acetate (0.46 mJ) it was likely that a spark of that energy would have ignited the vapour/air mixture. Then once the fire had started the powder/air mixture would have been ignited.

*Suggested safety measures*

- Ensure that the metal drum is grounded, either through the workman when he is wearing conducting footwear and is standing on a grounded surface, or directly by means of a grounding wire.
- The resistance between the drum and ground should not exceed 100 MΩ.

## 8.13 Hybrid mixtures

## 8.14 Grinding solvent-wet plastic

A plastic material, which had been wetted with toluene, was fed into a machine ready for grinding. The result was a quantity of fine particles of plastic enveloped in an atmosphere of solvent vapour which then fell through the bottom of the apparatus into an enamel coated steel vessel. As the grinding of the batch was nearing completion a fire suddenly occurred inside the machine.

Initially, investigations were focused on the hot surfaces

produced by friction during the grinding of the plastic as a possible source of ignition. Such sources are identified by the colours of the hot surfaces caused by annealing and by scratch marks caused by foreign bodies. No such coloration or scratch marks were found. Electrostatic discharges were not suspected because the grinder and the enamelled vessel were each grounded separately, the grinder via its electric cable and the vessel through the conducting floor on which it rested. With the exception of the wetted plastic there was no other plastic on the equipment.

Electrical tests on the vessel (volume 50 l) showed that the resistance between the vessel and ground was about 1.0 TΩ. One of the contact points for the resistance tests was a rusty area on the upper rim of the vessel where some of the enamel had been chipped off. We were surprised to find such a high resistance to ground considering the large area of the bottom of the vessel. However, on closer examination we discovered that the shape of the bottom was concave on the outside. This meant that only the periphery of the base made contact with the conducting floor. This restriction in the area of contact between the vessel and the floor would lead to a much higher resistance to ground than would be the case if the bottom of the vessel was flat, particularly as the vessel was coated with enamel.

To avoid any spillage of the product the grinder was fixed tightly onto the vessel. However, because of the enamel coating, the electrical resistance between the vessel and the grinder was very high. It was also noticed that at the place where the enamel was chipped there was a small indentation which, presumably, was caused by the blow which removed the enamel. This dent in the vessel would have provided a gap between the vessel and the grinder across which a spark could pass.

By checking the electric field locally our expectation that the wetted plastic had become electrostatically charged during grinding was confirmed. As the charged plastic filled the vessel charge was induced on the latter until its potential was sufficient to cause a spark at the gap in the presence of a hybrid mixture of toluene vapour and plastic particles.

*Suggested safety measures*

- Enamelled vessels must be reliably grounded when used for collecting electrostatically charged materials.

## 8.15 Rotating beater dryer (2)

A rotating beater dryer was being used to dry an organic product which was wetted with a flammable solvent (see 8.2). Although the machine ran satisfactorily, now and again the people in quality control complained of the presence of black particles in the product. It was assumed that the blackening occurred in the rotating beater dryer as it was known that fires might be started in that kind of apparatus. But this suspicion could not be valid as the drying was done under vacuum where combustion would be impossible because of the absence of oxygen. Being at a loss the plant manager tried to lay the blame for the blackening at the door of the expert on electrostatics. As we knew that any kind of incendiary gas discharge emits radio-frequency signals (see 3.3.2) we arranged for an AM radio receiver to be installed in the region of the feed opening. Contrary to the previous experience (see 8.2) only a few signals, at a rate of less than one per minute, were recorded. After some thought we remembered that the radio signals would only be detected in the presence of atmospheric pressure. Under vacuum, electrostatic discharges are of the glow discharge type, which are without the 'pinch effect' (see 3.3.2). This directed us to an examination of the vacuum pump system. After the dryer had been evacuated only solvent vapour would be extracted, so that the flow rate through the pump would fall to a low value. On installing a flow meter in the suction line we were surprised to discover that the expected drop in the flow rate did not occur, indicating that there was a leak in the system. We used smoke signal cartridges to locate precisely where the leakage was. Smoke was found being sucked into the bearing box, on the driving side of the rotating shaft, indicating a leak. Thus, it was possible that sufficient oxygen could, at least temporarily, be present for the combustion of the solvent vapour/dust mixture in the region of the loose bearing box. This was confirmed when it was found that the surrounding parts of the shaft entrance were blackened with soot. As combustion was restricted to that small region of the dryer, explosions could not develop and the incidents remained unnoticed except for the occasional presence of black particles in the product. As with the case previously discussed (see 8.2) the source of the ignitions was very likely sparks from the partially encrusted beating bars.

*Suggested safety measures*

- Because the gasket on the gearing box could not be reliably tightened it was suggested that a rig be set up to allow the introduction of nitrogen into the dryer, thereby preventing the access of air via the loose bearing box.

## 8.16 Shovelling solvent-wet powder

For several years a solid powder wetted with toluene had been delivered in galvanized steel drums to a plant for drying. It had been shovelled out of the drum into the funnel of a dryer. In view of the danger of electrostatic sparks all conductors were reliably grounded. The drum stood on a conducting floor and the worker wore conducting shoes and gloves.

The accident happened during the first shift after the weekend. A worker was shovelling out the product when suddenly the drum burst into flames. There was a combustible mixture of toluene vapour and dust in the drum which had ignited. A spark from the shovel, consisting of a steel blade and a wooden handle, was at first suspected as the source of the ignition. It was thought that the steel blade could have caused a mechanical spark when hitting the steel drum. Because of this it was considered that the steel blade should be replaced by one made of bronze. Although it is easy to blame ignitions on mechanical sparks, tests have shown that it is by no means easy to ignite solvent vapours with them.

The fact that it was early spring time, when the relative humidity of the air was low (30 per cent) and the plant was well heated, we felt that the accident could readily have been caused by an electrostatic spark. In general, wood is a moderately good conductor of static under normal relative humidity conditions ($>50$ per cent) but ceases to be so in warm, dry conditions. This could mean that the blade of the shovel was electrically insulated from earth by the wooden handle and, therefore, capable of delivering a spark when charged. Tests were done with a shovel which had been conditioned for three days in a similarly dry atmosphere. With the worker holding the shovel the electrical resistance between the blade and earth was found to be $40\,G\Omega$! With the shovel held inside the drum, but not touching it, the capacitance between the blade and the drum was about $30\,pF$. In another test the shovel was inserted into a drum filled with toluene-wetted powder, and on removing the shovel a potential of 6 kV was measured on the blade which remained for a short time. The corresponding maximum energy of the charge on the blade, calculated using the energy equation, was about 0.5 mJ. This would be dissipated to earth at a rate determined by the time constant of the system, i.e. 1.2 s. As the MIE of toluene is about 0.2 mJ, the energy of a possible spark between the shovel and the drum would be more than sufficient to ignite the solvent vapour. Because the time between the charging of the blade and the passing of the spark must be very short, in view of the small time constant, the worker would have had to move the shovel quickly as it was making contact with the drum. He must have been lively enough after the weekend to be able to do this, thereby causing an incendiary spark discharge.

*Suggested safety measures*

- It is unwise to rely on the conductivity of wood in all situations. A shovel with a conducting handle of, say, metal should be used and the operator reliably grounded.

## 8.17 Liquids

## 8.18 Emptying a drum via a glass pipe

This is a case in which toluene was being drawn from a metal drum by means of a vacuum pump, via a glass pipe, into an intermediate glass vessel of volume 40 l. The glass pipe and the vessel were joined by a 50 mm length of flexible PTFE tubing. When the drum was almost empty and the worker was removing the pipe from the drum there was a sudden loud bang! Afterwards the top and bottom of the drum was seen to be bulging outwards and the worker suffered a hearing trauma. The plant manager, who was not interested in a thorough investigation, immediately blamed an electrostatic discharge as the cause of the accident and called us in to confirm his diagnosis.

At first we could not support his suspicion because the metal drum was reliably grounded by a clamp. The worker had worn conducting shoes and the resistance through the floor and ground was less than 100 MΩ. The glass pipe was of 25 mm diameter, which is within the permitted limit of 30 mm laid down in the German Guidelines (see 3.6, (3)). Further, the area of the PTFE tube was about 13 $cm^2$ which is well inside the limit of 100 $cm^2$ permitted in the same guidelines. Even if it was assumed that air bubbles could have developed in the toluene, leading to the formation of a high electrostatic charge by the end of the emptying process, there was no hint of an electrostatic ignition under these circumstances.

On examining the glass pipe more closely we noticed a colourless solid inside it which covered almost the entire 1 m length of pipe. In the plant manager's statement this deposit was described as a precipitation of salt (potash). It originated from a mother lye which was passed through the pipe in a preceding step of the process. As it was not soluble in toluene it remained in the pipe as the toluene passed through it. The electrical resistance of this deposit along the full length of the pipe was about 10 kΩ. When the pipe was fully inserted into the drum the capacitance of the salt deposit relative to the grounded drum was 80 pF. The potential required on the deposit for it to store enough energy to ignite a sensitive mixture of toluene and air (MIE = 0.2 mJ) would be almost 3 kV. However, this deduction was not conclusive because as the glass pipe is pulled out of the drum the capacitance of the deposit decreases. Thus, for a given charge on the deposit there is a corresponding increase in the potential. So, according to the energy equation $W = 0.5CU^2$ the energy $W$ of the charge on the deposit increases by the same factor by which the capacitance $C$ is decreased, e.g. if the capacitance is halved, the energy is doubled. During the removal of the pipe from the drum a spark discharge could have occurred between the deposit and the bunghole of the drum as the end of the pipe came close to it. The spark could have ignited an explosible mixture of toluene and air.

No further investigations were deemed necessary in view of the fact that toluene is a chargeable liquid and the undoubted charging of the salt deposit. The formation of static charges has always to be taken into consideration with liquids of this type when they are flowing through pipes.

*Suggested safety measures*

- When handling non-conducting (chargeable) materials under

the regulatory conditions mentioned above it must be ensured that no conducting deposits can occur.

- In the case of switch loading it is recommended that only conducting and grounded equipment be used.

## 8.19 Funnel with a Mucon outlet

An organic powder which was delivered in 200 l metal drums had been poured into an agitator vessel containing toluene at room temperature. To avoid the spread of dust the lids of the drums had been replaced by a metal funnel which was attached to the drums by a metal clamping ring. A Mucon outlet was fitted to the narrow end of the funnel. By using a fork-lift truck the drum could be lifted up and turned upside down with the Mucon outlet positioned above the opening of the agitator vessel. The Mucon outlet was then opened manually allowing the powder to pour into the vessel and the displaced toluene vapour to pass into the drum. Thus, the workers were exposed neither to the organic dust nor the toluene vapour. With the danger of an electrostatic spark in mind, the funnel was grounded via the metal drum and the fork-lift truck, which had conducting tyres and was standing on a conducting floor. This procedure worked very well for a long time until, one day, we were informed that an explosion had occurred.

We could not believe that static would have been the cause of the accident. It was understood that all fork-lift trucks which are used in the presence of an explosible atmosphere have to be provided with conducting tyres; this had been checked and documented. However, when we inspected the site of the accident we found the burned remains of a plastic drum! In the light of this information we were immediately able to explain the sequence of events leading up to the accident. Because the funnel had been clamped by a metal ring to the plastic drum containing the powder it was, effectively, isolated from ground by the drum. The Mucon outlet was made of a synthetic fabric, so that a spark discharge from there was virtually impossible.

In view of the amount of damage caused by the accident we had to reconstruct the working conditions in order to quantify the possible size of the spark which could have been produced.

The funnel was clamped onto a plastic drum containing the organic powder and the assembly was lifted into an upside down position above the agitator vessel. The latter had been purged with nitrogen gas. Measurements showed that the resistance between the funnel and ground was 0.5 TΩ, and that the capacitance of the funnel to ground was 80 pF. As the powder ran through the funnel audible spark discharges could be heard in the region of the Mucon outlet and the potential on the funnel reached a maximum value of 4.5 kV. Using these data the calculated energy of the charge on the funnel could reach about 0.8 mJ. The time constant of the charged system was 40 s. As the MIE of the toluene is 0.2 mJ it was, therefore, certain that a spark discharge was the cause of the accident. The fact that the change from metal to plastic drums had taken place several months before the accident implies that during that time the conditions corresponding to the danger triangle (Fig. 1.1) had not existed.

*Suggested safety measures*

- All conducting parts of the system must be grounded.
- When exchanging metal parts for plastic ones in hazardous conditions, account must be taken of the danger from static electricity.

## 8.20 Metal valve in a glass apparatus

A glass apparatus was being used for the distillation of heptane. Beneath the condenser was a 50 mm glass pipe inside of which a valve was mounted by means of metal flanges. An alert worker observed spark discharges between a metal bolt on the valve and a nearby metal label, and immediately informed his foreman. As it was an explosion-endangered area in which large amounts of flammable liquid were being processed it was compulsory that precautions against any source of ignition be taken.

There was no doubt that the sparks were caused by static electricity generated by the heptane as it flowed through the valve and the consequent induction of charge on the metal parts of the valve. Yet, this was surprising because there had been instructions that all metal flanges had to be grounded (see 8.21). The valve itself, being made of plastic and with only a small metal bolt, did not have to be grounded. It was hardly believable that the observed sparks could possibly come from the isolated bolt which would be of such low capacitance that any sparks from it would be barely visible, let alone capable of igniting a hydrocarbon/air mixture. However, on examining the valve more closely it became evident that the counter flange of the valve was made of aluminium and that in order to make it look better it had been painted in the same colour, black, as the plastic part of the valve. Other aluminium flanges were left unpainted and were recognizable as metal parts which had to be grounded. Obviously we were glad that no ignitions had occurred in this case but felt challenged to discover whether or not the observed sparks could cause a sensitive mixture of heptane and air to ignite. Accordingly we made the appropriate measurements which yielded the following results.

| | |
|---|---|
| Potential on the bolt just before a spark discharge | 10 kV |
| Capacitance of the metal parts of the valve | 20 pF |
| Calculated energy of a spark | 0.80 mJ |

As the MIE of heptane is about 0.30 mJ the above results indicated the possibility of an ignition.

The worker who noticed and reported the sparks is a good example of a person who cares about safety at his place of work.

*Suggested safety measures*

- All metal parts in glass apparatus which may become charged by the streaming effect of liquids must be grounded if their diameter is 50 mm or more.

## 8.21 Spark discharge from an isolated metal flange

Incendiary spark discharges from metal flanges in apparatus made of glass are extremely rare and in a 23-year period we have come across only one incident.

On a cold winter's day when the humidity was low, a distillation apparatus was being used to separate toluene from carbon tetrachloride. A leaking flange allowed a small amount of toluene to drip down the side of the glass pipe which was at a temperature of about 60°C. At this temperature the toluene evaporated forming a combustible mixture with the air. The specified volume resistivity of the toluene was about 4 GΩ m indicating that it was prone to static charging. There were several ungrounded metal flanges on the 50 mm diameter glass pipe, each with a capacitance to ground in the region of 20 pF. One of these was almost touching a steel part of the rig which was grounded.

The effect of the charged liquid flowing inside the pipe was to induce charges onto the metal flanges, causing the potential of the flanges to rise to 8 kV within 40 s of the liquid starting to flow. As a result of this a discharge occurred between the flange and the steel part of the rig causing an ignition of the toluene/air mixture.

The MIE of toluene is about 0.2 mJ. By using the energy equation it was shown that the energy stored on the flange was 0.6 mJ and, therefore, more than adequate to ignite the toluene/air mixture.

The infrequency of such an accident was attributed to the improbability of there being an atmosphere of appropriately low ignition energy in the region of the spark gap.

*Suggested safety measures*

- Ensure that metal flanges and any other isolated metal parts on glass apparatus are reliably grounded.

## 8.22 Rubber hose with a supporting helix

Toluene, which had been delivered to a plant in a 500 l steel drum, was being emptied via a delivery pump into a reactor. As Fig. 8.6 shows, this was done with a hand-held suction pipe connected to the pump by means of a rubber hose reinforced with a steel helix (see insets). For easier handling and to prevent mechanical sparks we were asked if the suction pipe could be made of polypropylene. It was agreed that a plastic pipe with a maximum diameter of 30 mm would be permissible.

This handling of toluene had been carried out for several months before we were told that a severe explosion had occurred. The drum was torn apart and a worker suffered fatal injuries. What had happened? The procedure was to put the steel drum onto a steel floor with its bunghole uppermost. The suction pipe was inserted into the hole and the toluene was drawn out by the pump. When the drum was almost empty and some air had been

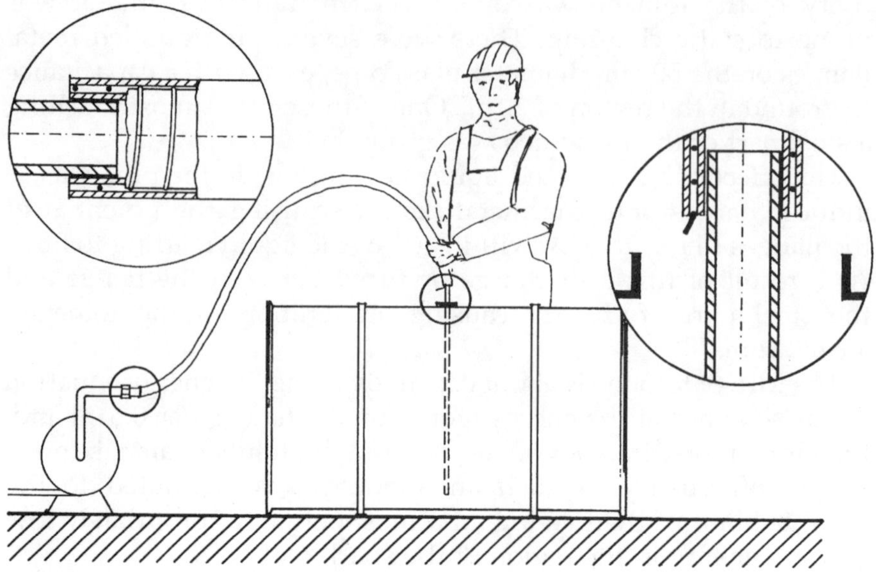

**Figure 8.6** Emptying a drum containing toluene by means of a suction pump

sucked into the pipe, the worker put the drum at an angle to get at the remaining toluene. To do this he had to insert the suction pipe as deeply as possible, being limited only by the rubber hose which was attached to the pipe's elbow. It could be demonstrated that it was at this very moment that the explosion occurred. Those in charge first suspected the plastic pipe as having caused the ignition. But this could not be true as such small pipes were generally known to be safe. Then the reliability of the grounding was questioned because the grounding clamp was found not to be fixed to any pieces of the drum. But this did not matter because similar drums standing on the same floor had resistances to ground of below 1.0 MΩ in any position. Despite the evidence against static being the cause of the accident we were still unconvinced and a decision was made to pursue the matter further. Attention was now directed to the rubber hose. The rubber itself was insulating and the resistance between the steel helix and an inserted metal pipe was 1.0 TΩ. A test was done in which 10 l of the same toluene (resistivity 10 GΩ m) was fed through the hose.

This produced a potential of 400 V on the helix. On the basis of this result the cause of the accident became clear.

The toluene had become electrostatically charged as it flowed through the plastic pipe and the rubber hose. The electric field from the charge at the inner wall of the rubber hose had, by induction, produced a similar charge on the steel helix. The effect of this was to raise its potential to a level at which a spark could pass between the helix and any nearby conductor. When the worker inclined the drum and inserted the suction pipe into it until the steel helix almost touched the rim of the bunghole, a spark discharge took place. Doubtless there was a combustible toluene/air mixture close to the bunghole, which was ignited by the spark. As the pump was still causing air to pass into the drum, because of the displaced toluene, a combustible toluene/air mixture would be present inside the drum which, on being ignited, caused the drum to explode.

We were very concerned when we learned from a colleague that in another company a similar fatal accident had occurred under almost the same circumstances as those outlined above.

*Suggested safety measures*

- When handling combustible liquids use conducting rubber hoses only.
- Care should be taken to ground the rubber hose as well as the metal helix inside it.

## 8.23 Isolated steel spacer in a metal pipe

Although this case is a rarity, it is important that attention be drawn to it.

A worker reported that he had noticed a small tongue of flame at the flanges of an exhaust pipe system. This had to be investigated because it occurred in an area of the plant where the hazard of an explosion was possible. We were shown where the incident occurred and what we found can be explained by referring to Fig. 8.7. The flanges were on a steel pipe of diameter 100 mm which was enamelled on its inside surface. To bring this pipe into line with an agitator vessel, which was being ventilated, a steel spacer of length 40 mm was fixed between the two flanges.

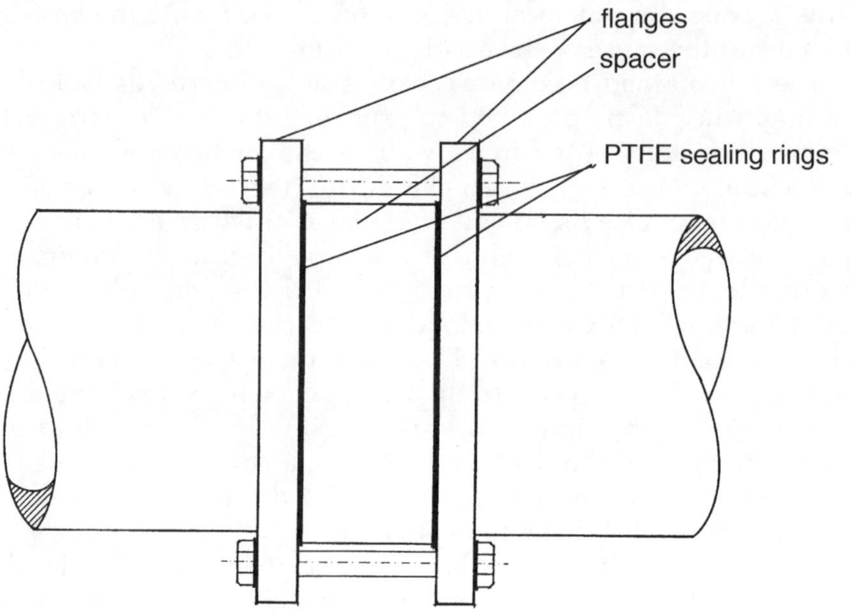


**Figure 8.7** Metal flanges

Sealing rings of PTFE were fitted at each end of the spacer and the assembly was secured with bolts. Sometimes, when there was a need for strong ventilation, e.g. when the manhole of the vessel was opened, sparks could be seen across a 2 mm gap between the spacer and a bolt. Plainly, static electricity was the cause of the sparks but an investigation was needed to discover how the ignitions occurred.

The electrical capacitance of the spacer relative to the grounded flange was found to be 28 pF. Tests for the potential on the spacer showed that values of at least 6 kV could be reached before a spark discharge occurred. The corresponding energy of the charge stored on the spacer could be 0.5 mJ or more. The resistance between the spacer and ground was 6 TΩ and the charge decay time was 168 s. As the various hydrocarbons which were transported in the pipe were of MIEs between 0.2 and 0.3 mJ they could easily have been ignited by a spark from the spacer. However, a few questions still remained to be answered.

It is well known that gases flowing in pipes do not, themselves, cause static charging (see 2.2.2). As has already been mentioned, sparks were only observed when the manhole of the vessel was opened (or closed). The resulting changes in temperature and pressure in the system could easily cause the formation of aerosols which are capable of becoming electrostatically charged. Such charges, when stored on the spacer, could then raise the potential of the latter to a level at which an electrical breakdown between the spacer and a bolt would occur. Because of surges of pressure in the pipe, hydrocarbon vapours were liable to escape through the PTFE sealing rings for short periods of time. As the coincidence of a spark discharge and the leakage of the vapour would be a rare event, so ignitions of the vapour would be expected to be very infrequent.

Spark discharges between pairs of metal rings inserted between PTFE seals have been reported to us but without any ignitions.

*Suggested safety measures*

- The grounding of isolated metal parts which might become electrostatically charged is, of course, compulsory in explosion-endangered areas. However, implementing such measures on isolated metal objects appears to be difficult and troublesome.

## 8.24 Filling a metal drum on mobile scales

A special request was made by a customer for 40 kg of an intermediate product dissolved in hexane which normally was delivered in 1 $m^3$ IBCs mounted on wooden pallets. The foreman put a 60 l steel drum onto some metal scales and by means of a hand pump and a conducting hose transferred the mixture from the IBC to the drum (Fig. 8.8). He was familiar with the problems that can arise from static charges and, in particular, that hexane is a chargeable liquid. As he was wearing conducting shoes and knew that the floor was conducting he felt quite safe. It was, therefore, to his great surprise when at the finish of transferring the product a flame shot out of the drum which ignited the liquid inside. He quickly stopped pumping the liquid, took hold of a carbon dioxide extinguisher and put out the fire. Because he had taken all the necessary safety measures he subconsciously thought that static electricity had played a trick on him. He was at a loss

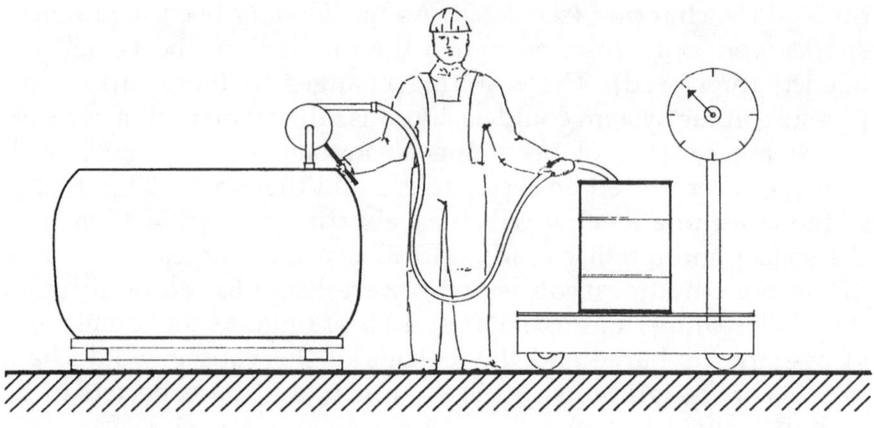

**Figure 8.8** Filling a metal drum resting on mobile scales

to know how the fire had started. He reported the incident to the manager who also could give no explanation. Feeling rather embarrassed they went sheepishly to the electrostatics expert and quietly asked his advice on the matter. The only test that the expert did was to measure the resistance between the steel drum and ground via the metal scales. Surprisingly the value was greater than $1.0\,\mathrm{T}\Omega$. What could have been the cause of this high resistance? Apparently it was due to the solid tyres of the scales which were of a highly insulating rubber! The tyres were originally of an electrically conducting rubber, but because they had become swollen through the spillage of hexane they had been replaced by a solvent-resistant type which, unfortunately, was electrically insulating. Thus the charge induced on the drum (and the scales) from that on the hexane inside the drum could not rapidly escape to earth. But how did the spark discharge occur? Generally it should be expected that the conducting hose would make contact with the rim of the drum, thereby allowing the charge to be dissipated to ground via the IBC and its wooden pallet. However, in this case the foreman took great care not to touch the drum with the hose in order to avoid disturbing the scales. When he had almost pumped in 40 kg of the mixture the metal nozzle of the hose came close enough to the rim to cause a spark discharge.

*Suggested safety measures*

- All conducting parts of the equipment must be reliably grounded.
- If conducting tyres are required only those which are appropriately marked should be used.

## 8.25 Slicing solvent-wet plastic

An elastomer, wetted with hexane, had to be sliced in a shredding machine in the laboratory. This had been done many times before, the slices being collected in a PE bag attached to the outlet of the shredder. A laboratory technician, who knew something about electrostatics, remembered that PE can be hazardous especially in the presence of a combustible solvent vapour atmosphere. He, therefore, decided to replace the PE bag with a steel drum resting on a wooden stool with its legs shortened so that it fitted perfectly under the shredding machine, as is shown in Fig. 8.9. When everything was in place and the laboratory staff had assembled to watch the test run the shredder was switched on. Shortly after the start a tongue of flame shot out of the drum which, luckily, did not cause any severe damage.

The staff were, no doubt, confused that after taking the precautions to avoid a potential ignition hazard a fire had occurred. Unable to conceal his anger the head of the laboratory sent for the electrostatics expert who made a number of measurements to establish the cause of the fire and to show the staff how it had happened. The results were as follows:

| | |
|---|---|
| Resistance $R$ between the drum and ground | $0.5\,\text{T}\Omega$ |
| Capacitance $C$ between the drum and ground | $80\,\text{pF}$ |
| Distance $d$ between the drum and shredding machine | 2–10 mm |
| MIE $W$ of the hexane/air atmosphere | $0.2\,\text{mJ}$ |
| Volume resistivity $\rho$ of the polymer | $1.0\,\text{T}\Omega\,\text{m}$ |

The resistance between the drum and ground, via the wooden stool, was unexpectedly high and was caused by a coating of synthetic resin on the stool. Using the data given above the following parameters were calculated.

Time constant of charge decay $\tau = RC$ is 40 s
Potential on the drum $U = \surd(2W/C)$ is 2.2 kV

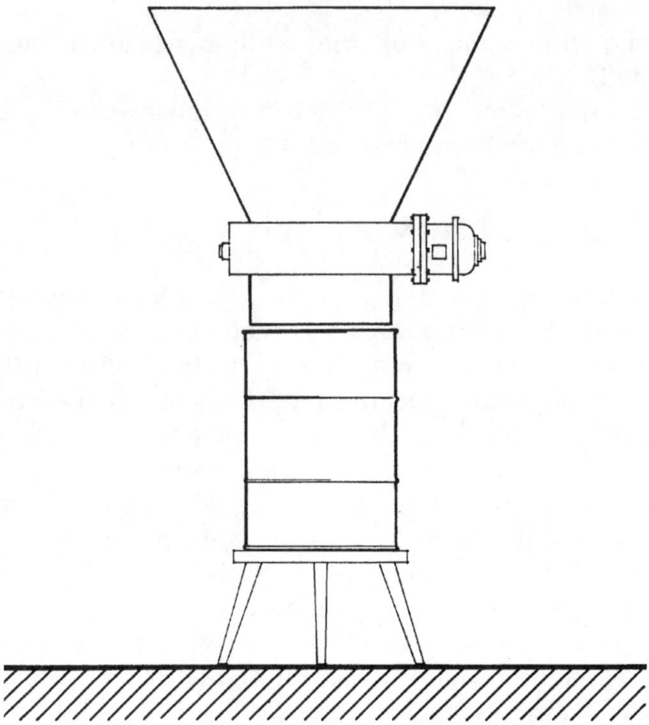

**Figure 8.9** Steel drum resting on a wooden stool underneath a shredding machine

From experience it is known that materials with volume resistivities similar to that quoted above can become highly charged. The field from the charged material inside the drum could, by induction, have raised the potential of the drum to 2.2 kV or more. By this means a spark sufficient to ignite the hexane/air mixture could have passed between the drum and the grounded machine.

*Suggested safety measures*

- Ensure that metal containers are reliably grounded.
- Bare wood is electrostatically dissipative in normal atmospheric conditions. However, it can become insulating after being coated with certain resins and varnishes. When in doubt make the appropriate electrical resistance tests.

## 8.26 Application of rubber adhesive

Rubber adhesive is a solution of india-rubber in benzene. It is not only used for repairing punctured bicycle tyres but also as an adhesive for sealing steel vessels. In Fig. 8.10 is shown a steel silo of 2.5 m diameter in a horizontal position. A worker was applying the adhesive from a hand-held metal bucket to the inside of the silo with a roller brush. Although the vessel was equipped with a ventilation system (not shown) to provide the worker with fresh air the possible danger of an explosion within the silo could not be ruled out. Precautions had already been taken to avoid any ignition sources. An independent air-cooled electric light had been installed. Measures were also taken to avoid an electrostatic hazard by the use of a roller brush with a conducting handle, by providing the worker with conducting gloves and by attaching a grounded cable to the worker's wrist. The latter was necessary because the worker had to wear soft plastic galoshes to protect the already finished coating on which he stood. This had all been properly organized and we felt that everything had been made as

**Figure 8.10** Rubber adhesive being applied to the inside of a steel vessel

safe as possible. We were, therefore, amazed on being informed that a fire had occurred while the worker was applying the adhesive. Fortunately he was not severely burned as he was able to escape quickly through the manhole. On first inspection it appeared that all the agreed safety measures had been applied. However, during tests we found that the resistance between the wriststrap and ground was greater than 1.0 TΩ, although the cable and its grounding screw looked to be in perfect condition. Only by flexing the PVC coated copper braid did we find the latter to be broken while the coating remained undamaged. In tests where this type of customary cable was put under tension it was found that the copper braid tore apart without any visible damage to the PVC coating. For proof that the accident had been caused by an electrostatic spark it was insufficient to show that the worker had not been grounded. In addition it had to be established how and to what degree a charge had been produced. On reconstructing the situation we were surprised to discover that when a person, who was isolated from ground, moved the roller brush up and down the silo surface ten times a potential of 16 kV was produced on the person! This is well in excess of the several kilovolts that are required on a person of average capacitance to cause the ignition of a benzene/air mixture by a spark discharge from the body. In the present case the spark could have occurred at several places, but was most likely to have happened between the hand-held metal bucket and the wall of the silo.

*Suggested safety measures*

- Ground connections should be tested regularly using a megohmmeter, as visual inspection of leads and contacts can be misleading.

## 8.27 Valve with a corroded PTFE coating

An explosion occurred when an agitator vessel, partly filled with toluene, was loaded with powder from a metal drum containing a PE liner. For obvious reasons static electricity was suspected as being the cause of the ignition. Investigations were carried out which led to the following results. Assuming that an explosive toluene/air mixture was present near the vessel opening, two electrostatic sources of ignition were possible. The manually operated steel lifting device, which was on insulating rollers, was used to lift the drums above the vessel opening and empty them by turning them over. An incendiary spark might have occurred between the isolated drum and the vessel opening. At the end of the emptying procedure the PE liner, together with the remaining product, was taken out of the drum by hand and was shaken out into the agitator vessel. An incendiary brush discharge might have occurred between the charged PE liner and the vessel opening.

The worker involved was lucky not to have been burned when the tongue of flame leaped from the vessel opening. As he stated, he had already quickly stepped aside and was busy folding the empty PE liner. The company could well be pleased with this agile worker but we were not satisfied with the idea of such a slow moving flame.

What was really going on behind the scenes?

Above the agitator vessel there was a glass exhaust pipe which attracted our attention because its inside was blackened with soot. The pipe was attached to the top of the agitator vessel by means of a porcelain valve. A routine check showed that the metal parts of the valve (casing and spindle) were grounded. Further examination of the valve revealed that a PTFE lid for the seal was coloured brown whereas, as expected, all other PTFE parts were white. The brown colouring was obviously caused by corrosion. When checking the conducting behaviour of the coloured lid we found that its resistance was about 1.0 MΩ. The resistance of the white PTFE, in which the lid was embedded, was greater than 10 PΩ. The capacitance to ground of the coloured lid was found to be 12 pF. These data indicated that for a spark from the lid to ignite a toluene/air mixture (MIE = 0.2 mJ) the potential on it would need to be at least 6.0 kV.

When filling the agitator vessel with powder the emerging fine particles of dust were sucked out through the glass exhaust pipe via the valve. The collision of many dust particles with the lid would cause the latter to become highly charged. As the lid was isolated from earth the charge on it would raise its potential to a high value. Tests in the laboratory showed that a potential of 15 kV was reached on the lid, corresponding to an energy of five times that required for the ignition of a toluene/air mixture. Looking at the accident in this light it was deduced that the following sequence of events was likely. When the worker shook the PE liner a large number of particles were set free. These were sucked through the valve where they gradually caused a rise in the potential of the lid until a spark discharge occurred. Because of the presence of the toluene/air mixture an ignition, which began at the valve, was propagated in two directions, one causing a blackening of the glass pipe and the other an explosion in the agitator vessel.

*Suggested safety measures*

- Use grounded valves coated with conducting PTFE.
- Use grounded metal drums without PE liners in the presence of combustible solvents.

## 8.28 Fire during a coating process

A PET-foil of 15 μm thickness was being coated at a speed of 3 m/s at a coating plant. The coating itself was a solution containing ethyl acetate which being combustible presented a fire hazard. Indeed, an ignition occurred causing a fire in the area between the application roller and the pre-dryer. In such a situation one would, justifiably, blame a static discharge as the cause without considering any other ignition source. In particular, a brush discharge would be suspected. As has already been pointed out in 3.3.2, brush discharges are only capable of igniting sensitive combustible vapour/air and gas/air mixtures. When investigating situations where brush discharges have occurred experience has shown that it is rare for an optimally combustible atmosphere to be present for most of the time. Therefore, in cases like this one, we should not be content with assuming that a brush discharge was the only ignition source. We should look for other sources, e.g. propagating brush discharges and spark discharges. Each of these types of discharge can release higher amounts of energy than do brush discharges and can ignite less sensitive mixtures. On inspecting the machine we found various ungrounded metal parts in the vicinity of the application roller. Among these was a metal rod mounted across the foil and in front of the application roller. Measurements on the rod showed that it had a capacitance to earth of 200 pF and a resistance to earth of about 10 TΩ.

Although the rod had no contact with the highly charged moving foil, tests showed that its potential could reach 5 kV due to the effect of induction from charge on the foil. The likely discharge, therefore, was from the rod to a nearby grounded or large conductor in the solvent/vapour atmosphere. According to the energy equation the energy released in a single discharge could be as high as 2.5 mJ. This is about five times the MIE of the solvent/air mixture.

*Suggested safety measures*

- Ensure that all metal parts in the vicinity of the combustible atmosphere are reliably grounded. This applies to parts that can be charged by contact electrification and by induction.
- Be aware of the ignition hazards caused by brush discharges.

## 8.29 Polyvinyl chloride (PVC) hose partially wetted with water

In a degassing procedure butadiene vapour was being drawn off a caoutchouc polymerization reactor by means of a vacuum pump. A hose of length 0.8 m and diameter 40 mm, which was attached to the suction pipe on the pump, was made of transparent PVC in order that any plugs of water which might be carried along with the vapour could be seen.

A fire occurred in the region of the hose and later it was found that the hose had come away from the pump. Undoubtedly it was a butadiene/air mixture that had been ignited but the source of the ignition was unknown. In their investigations the plant management thought that static electricity was unlikely to be the cause of the fire because the inside of the hose was always wet due to the transportation of small quantities of water through it. In the end an expert on static electricity had to be called in to discover the source of the ignition. Tests were done on a similar length of old PVC hose, in a dry condition, in the laboratory. The surface resistivity of the inside of the hose was found to be 3.0 TΩ and the resistance of the entire length was 6.0 TΩ. These values indicated that the PVC was just about chargeable but there was no evidence to show that it could be charged in a wet condition.

Consequently, it was decided to do some further tests. The PVC hose was suspended by nylon threads in an inclined position and a metal pipe of length 50 mm was pushed onto its lower end. An electrostatic voltmeter was connected to the metal pipe. Water was sprayed into the higher end of the hose until it dripped out through the metal pipe. The voltmeter indicated that there was no charge on the pipe. Then the hose was suspended in a horizontal position so that a small amount of water when passed into it remained inside the pipe. When compressed air was blown into the opposite end of the hose to that with the metal pipe the voltmeter indicated briefly a potential of about 6.0 kV. At first we found this behaviour unbelievable but further tests showed that it was reproducible. This convinced the plant manager and ourselves that even water when moved by pressurized air can cause charge separation. But how was the fire started?

The spasmodic transportation of water by a gas or vapour can partially charge water at the inner wall of the PVC hose as has been demonstrated above. When the hose slipped off the grounded suction pipe on the pump a spark discharge could have occurred between the end of the pipe and the charged water in the hose. Generally, during degassing there is no explosible atmosphere in the PVC hose because of the absence of oxygen. However, at the instant the hose slipped off the pipe oxygen from the air would mix with the butadiene vapour at the place where the expected spark discharge was released.

*Suggested safety measures*

- The PVC hose should be replaced by a metal pipe in which is installed a glass viewing window.

## 8.30 Glass vessel containing a mixture of toluene and water

Toluene containing a small amount of water had been pumped from a drum into a 100 l intermediate glass container. On passing the liquid from the container into a steel agitator vessel situated underneath, an explosion took place just as the container became empty. Because of the presence of air in the glass vessel an explosible mixture with toluene vapour could have developed. The plant manager was able to exclude any uncontrolled chemical reactions and as no other ignition sources were present suspicion fell on an electrostatic discharge as the probable cause of the accident. This seemed quite plausible to us because with a mixture of toluene and water there would be a large interfacial area of contact between the components where charge separation could occur (see 2.2.1). However, two questions needed to be answered. First, where was the charge in the discharge stored initially and, second, how did it come to cause an incendiary spark discharge? As the glass vessel was shattered in the accident a full-scale laboratory test was done using another vessel containing a similar toluene/air mixture as that used before. It had been discovered that the original mixture contained traces of common salt in the water. The volume resistivity of the liquid was found to be 0.5 TΩ m. This figure classifies the liquid as being static prone. Measurements of the field strength taken near the jet as the liquid was released from the vessel revealed values of about 100 kV/m. When the glass vessel was empty we noticed that on the inside the lower part was covered with a film of liquid. The liquid was, obviously, water and had a surface resistivity in the region of 1.0 MΩ. This film of water acted as a conducting island which was insulated from ground by a resistance of between 0.1 TΩ and 1.0 TΩ.

Knowing that the outlet of the original glass vessel was connected to the steel agitator vessel by a short flexible tube of PTFE, we were able to deduce the course of events which led to the accident as follows. Some of the charge in the water component generated by the pumping of the mixture remained in the film of water when the mixture was emptied out of the glass vessel. As the breakdown strength of toluene is much higher than that of air the potential of the film of water, when covered with toluene, could rise to a far higher value than would be possible with the film in contact with air. Thus, at the instant when the last of the mixture ran out of the vessel an electrical breakdown above the film could have occurred causing the mixture of toluene vapour and air to ignite.

*Suggested safety measures*

- When handling two-phase systems in which the insulating phase is the continuous one, the system should be purged with an inert gas.

## 8.31 Person wearing protective gloves

When pumping ethyl acetate from a metal drum via a plastic hose into an open stainless steel tank of volume $1\,m^3$ an ignition occurred which was followed by a fire. As ethyl acetate is classified as a non-chargeable liquid because of its low volume resistivity ($40\,M\Omega\,m$), no attention had been paid to safeguarding against an electrostatic hazard. As all other possible ignition sources could be excluded by the plant manager the question of static electricity was raised and the appropriate tests were made. The plastic hose, which was made of silicone rubber, showed a resistance of more than $100\,T\Omega/m$. The hose had a diameter of 25 mm (German Guidelines (see 3.6 (3)) permit up to 30 mm) and, for ease of handling, was fitted with a stainless steel nozzle at its outlet. The worker wore latex protective gloves which had a resistance through them in excess of $1.0\,T\Omega$. His shoes were of a resistance of about $100\,M\Omega$ and the resistance of the floor was in the region of $1.0\,G\Omega$. As this latter value is above the acceptable limit of $100\,M\Omega$, tests were done of the resistance to ground of a person wearing conducting shoes and standing on the floor. The result was a resistance to ground of rather less than $100\,M\Omega$.

The metal drum, pump and steel tank were each reliably grounded. The only conducting part which was not grounded was the steel nozzle. When the nozzle was held by a grounded worker wearing the latex gloves its resistance to ground was about $1.0\,T\Omega$ and its capacitance was 100 pF. But the question remaining was how could the nozzle have become electrostatically charged? Fortunately, the worker in charge of the transfer of the liquid was only slightly injured and he told us that the ignition had happened when the drum was nearly empty. At that stage the pump would be removing the liquid only spasmodically. Thus, the steady flow of liquid needed for charge equalization would no longer apply (see 2.2.1). Under these conditions even conducting liquids can give rise to static electrification.

The course of events were reconstructed, while taking the appropriate safety measures, and it was found that the potential of the nozzle when held in the gloved hand quickly reached 3.5 kV. Using the energy equation the calculated energy of the charge on the nozzle at this potential was 0.6 mJ, which could easily cause the ignition of a sensitive mixture of ethyl acetate and air (MIE = 0.46 mJ). This test clearly indicated that static could have been the cause of the ignition. In this case it was not difficult to deduce where the spark discharge occurred. It was where the nozzle approached the wall of the tank in which there was an explosive mixture.

*Suggested safety measures*

- In hazardous areas all metal parts should be grounded, even when it is not apparent that static electrification can occur.

## 8.32 Running off ethylene oxide into a plastic bucket

A tank of liquefied ethylene oxide was being emptied of the few remaining litres for disposal. Because the tank was under a low positive pressure the liquid was run off through a valve at the bottom of the tank into a plastic bucket of volume 10 l. To prevent too much evaporation of the liquid some water had previously been added to the bucket which floated on top of the liquid. According to a workman an ignition occurred immediately after the valve was opened.

As the temperature of the vessel was about 16°C and the boiling point of ethylene oxide is about 12°C the presence of a combustible mixture of vapour and air was to be expected. The workman was wearing conducting footwear and was standing on a floor of conducting concrete. His resistance to ground was less than 20 MΩ which indicated that there was no possibility of a spark discharge from his body.

Even so, our assumption that the ignition was caused by an electrostatic discharge became plausible when we were told that, in this case, the bucket had been taken from a pile of new ones stored on top of each other. On reconstructing the incident we found that a new bucket taken from the pile was highly charged and the associated electric field was about 800 kV/m. The charge was produced during the removal of the bucket from the pile, as measurements showed a similar charge of opposite polarity on the next bucket in the pile.

After pouring 3 l of water from another (old) plastic bucket into the charged bucket a potential of 6 kV was measured between the surface of the water and ground by using a static voltmeter and an insulated probe immersed in the water. The electrical capacitance between the water, in the hand-held bucket, and ground was about 30 pF. By substituting these data into the energy equation ($W = 0.5CU^2$), where $W$ is the energy, $C$ the capacitance and $U$ the voltage, the stored energy was found to be 0.5 mJ. The MIE of ethylene oxide is 0.07 mJ and its volume resistivity is 0.1 MΩ m. Using this information we deduced that the following was the likely course of the accident.

The jet of ethylene oxide leaving the grounded tank provided a conducting path to the tank. As the jet approached the charged surface of the conducting water a spark discharge occurred between the two liquids in which was released sufficient of the stored energy to ignite a combustible vapour/air mixture in that region.

*Suggested safety measures*

- Replace the plastic buckets with metal ones. The grounding of such buckets is compulsory and is achieved through the grounded workman. If he is wearing gloves they should be of a conducting material.

# Chapter 9
# Case histories not related to static electricity

## 9.1 Pouring powder into oleum

A deflagration occurred when an intermediate dye-stuff product of the aminobenzene group (A) was poured into an agitator vessel from a drum containing a PE liner (Fig. 9.1). The product was to be sulphated with oleum contained in the vessel. After the second drum was emptied a bright yellow tongue of flame was seen coming out of the vessel opening. As it was well known that the product was combustible a static discharge was suspected as being the cause of the ignition, particularly in view of the presence of the PE liner. We were given every facility to find out whether or not a static discharge really was the cause of the fire. To begin with we measured the MIE energy of the product/air mixture which was found to be 15 mJ. Although this is a moderately low value we doubted whether it would be achievable by a discharge from the liner. We then examined the grounding of the various parts of the equipment, e.g. the funnel, drum and the fork-lift truck itself. Although we were satisfied that the funnel was grounded at the time of the accident this could not be said, unequivocally, of the truck. The floor was electrostatically dissipative, but not the tyres of the truck. The resistance between the drum and the truck was of the order of megohms and would easily allow charge to pass

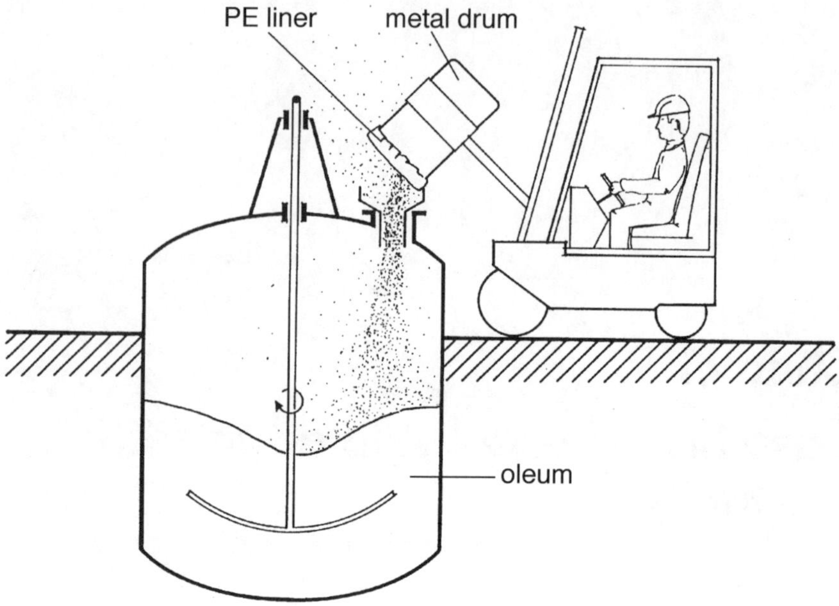


**Figure 9.1** Pouring powder into an agitator vessel containing oleum

between the two. Assuming that charge was generated during the separation of the product from the liner, the emptying procedure was repeated under almost identical conditions to those in place at the time of the accident. To simulate the worst conditions the four tyres of the truck were insulated from ground by placing PE foil beneath each of them. Three drums were emptied one by one into the vessel and each time the potential on the truck was measured. The maximum voltage reached was 1.0 kV. The capacitance of the truck relative to ground was found to be 900 pF. Thus, the maximum energy available for a spark discharge from the truck would be about 0.4 mJ. We had good reasons for assuming that a spark could have occurred between the metal drum and the grounded steel funnel.

While the basic problem seemed to be solved we were concerned about the difference between the MIE needed to ignite the product and the energy of the spark found in our tests. The latter was only 2.7 per cent of the former. Even taking into account the notoriously poor reproducibility of electrostatic contact charging measurements we were still doubtful about the level of charge found when emptying the drum. Of course, the proposed safety measure was to ground the truck, e.g. by using conducting tyres, but we had to point out that this would not be a definitive solution. We noted, rather reluctantly, that the possibility of a chemical reaction could not be ruled out.

About six months later we were informed that an exothermic chemical reaction was possible between oleum (vapour phase) and the product.

We put the case on file.

## 9.2 Fire produced on draining off residual benzene into a plastic drum

The progress of the accident is discussed below with reference to Fig. 9.2.

A sediment of a charcoal/benzene slurry had settled at the bottom of a condensate reservoir. The temperature of the slurry was 60°C and it had to be drained off, as was the routine, through a stopcock into a drum. As there had been an undue delay since the last drainage the system had become blocked so that nothing came out when the stopcock was opened. By poking the blockage with a welding wire, which happened to be at hand, the worker after considerable effort managed to free the slurry. He reported that within a few seconds a flame appeared at the stopcock which ignited the flowing benzene. As the man was wearing the standard safety outfit (helmet, eye protection glasses and gloves) he escaped with no more than a fright. The fire brigade was called and did an excellent job of controlling the fire, thereby avoiding any serious damage.

Once everything had settled down the technical managers decided to investigate the cause of the fire on their own. They started systematically, as they had been taught, by going through the 13 possible ignition sources and eliminating them one by one. Some sources were ruled out from the start, e.g. lightning, electromagnetic waves, ultrasonics and adiabatic compression. Further, as the emptying of the reservoir was a purely mechanical procedure a chemical reaction was also out of the question. The

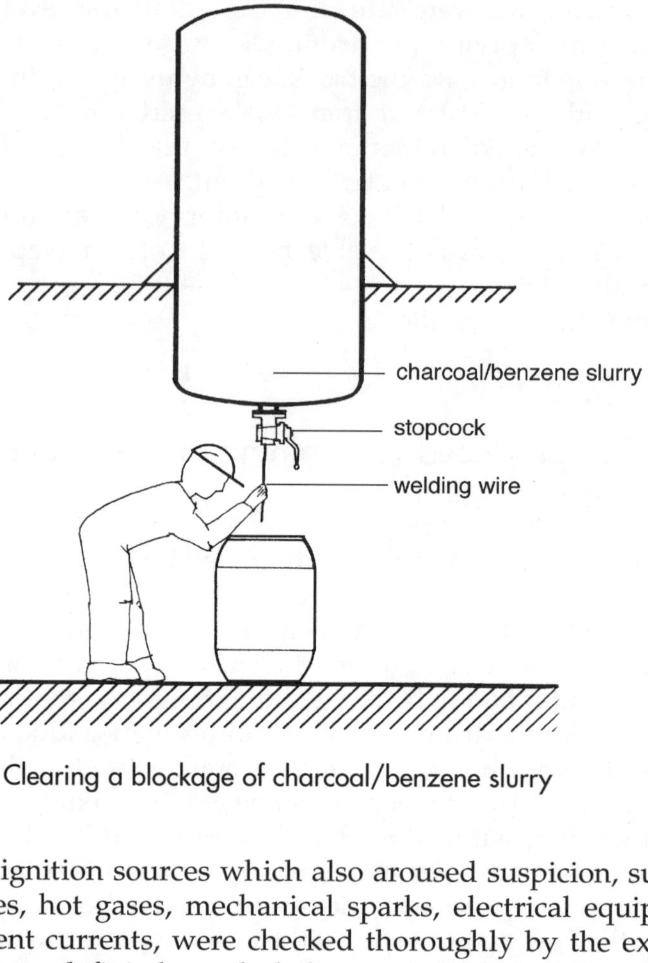


**Figure 9.2** Clearing a blockage of charcoal/benzene slurry

remaining ignition sources which also aroused suspicion, such as hot surfaces, hot gases, mechanical sparks, electrical equipment and transient currents, were checked thoroughly by the experts. Each one was definitely excluded as a possible ignition source. Even static discharges did not seem likely at first until they found that a plastic drum had been used and that it had been partially burned!

The frequency of emptying the reservoir was clearly stated in the operating instructions as was the requirement of a special metal drum, which had to be grounded, for collecting the slurry. Further checking showed that the recommended interval between emptying the reservoir had been considerably exceeded and that the worker, being unable to find the metal drum, had used a plastic one which happened to be lying around. Experienced managers would naturally tend to assume that accidents of this kind happen when the operating instructions have been disobeyed. It therefore seemed obvious to them that the improperly used plastic drum was the cause of the accident. Consequently, the electrostatics expert was asked simply to certify that, indeed, the plastic drum had been the cause of the ignition. Despite this request the expert was anxious to reconstruct the accident in order to discover the location of the charge and the spark. The resistivity of the slurry was found to be between 10 GΩ m and 100 GΩ m. It should be expected that the slurry would be charged when flowing out of the reservoir thus leading to a build-up of charge at the bottom of the drum. With this in mind the question arose as to how a charge at the bottom of the plastic drum could cause an incendiary gas discharge at the stopcock. Further investigation showed that the spark could not have come from the worker via the welding wire as he was wearing electrically dissipative shoes and gloves and was standing on a conducting floor which was grounded. A charge transfer arising from induction was also ruled out as all nearby conductors were grounded.

Critics were of the opinion that the worker's statement about the location of the ignition was wrong. They thought it took place at the bottom of the drum. But this would not be possible as there were no electrodes there to initiate a brush discharge. When an expert dares to refute the consensus view on the source of an ignition his colleagues expect that at least he will be asked indignantly what else it could be. To tell a chemist that an unknown chemical reaction might have caused the accident would seem to be an unforgivable stupidity. However, this suspicion was expressed in view of the slurry's excessive period of dwell in the reservoir.

Ultimately, a sample was taken from the charred remains of the slurry and subjected to DTA (differential thermal analysis). At 25°C the sample already showed an oxidation reaction which would cause an exothermic effect. It had to be concluded that the charcoal slurry in question, which was at a temperature of 60°C, was in a highly oxidizing condition and the resulting heat was sufficient to cause an auto-ignition of the slurry in air. There was every reason to believe that the thermal instability of the slurry was caused by the longer period of dwell in the reservoir.

## 9.3 Shovelling solvent-wet powder

In a suction filter made of porous stones in which the stirring was done by means of a titanium paddle, a dispersion of pharmaceutical powder in hexane had been partially dried under vacuum. A worker was busily shovelling out the hexane-moist powder. When he had almost finished and was attempting to scrape off a deposit of powder from the paddle with his shovel an ignition occurred which set the filter on fire. Static electricity was straight away blamed as the cause of the fire. A spark discharge from the shovel, which comprised a wooden handle and a carbon-steel blade, was suspected (see 8.16). Although this explanation seemed plausible we wanted the evidence to substantiate it. We first checked the dissipative resistances. The resistance between the shovel blade and the worker holding the wooden handle was between 1.0 GΩ and 10 GΩ. That between the worker and ground, through the shoes and the floor covering, was 0.5 GΩ. The capacitance of the shovel blade when held between two arms of the paddle was 27 pF. Knowing that the MIE of hexane is 0.2 mJ we calculated that the potential of the blade would need to be, at least, 4 kV in order to produce an incendiary spark discharge in the presence of a mixture of hexane vapour and air.

In reconstructing the event about 2 kg of hexane-moist powder was poured over the hand-held shovel several times while the blade was connected to an electrostatic voltmeter. In each test the potential on the blade did not exceed 100 V. This small value was consistent with that expected in view of the low dissipative resistance of the shovel given above.

Thus, the suspicion that an electrostatic discharge was the cause of the accident was proved to be unsustainable.

There remained the possibility of a mechanical spark as being the source of the fire. As no information was available at the time on the incendiary behaviour of mechanical sparks it was recommended that a blade of beryllium-bronze, a material which does not produce mechanical sparks, be used.

About two years later in another plant a fire occurred again under almost the same circumstances as the case discussed above. The blade of the shovel used was of carbon-steel. In view of the extremely high cost of beryllium-bronze and its poor durability, further enquiries were made with regard to mechanical sparks.

The results of work published by Bartknecht (1) have shown that striking rusty carbon-steel can cause sparks which will ignite hydrocarbon/air mixtures. However, stainless steel causes only glowing sparks which are not capable of igniting hydrocarbon/air mixtures.

*Suggested safety measures*

- Shovels with blades of carbon-steel, especially when rusty, should not be used in places where flammable gases or vapours are present.

## 9.4 Fire in an agitator vessel

A fire had occurred in an agitator vessel where casing head gasoline (flash point −20°C) was present, when sodium phenolate was fed into the vessel via a PE tube. Under normal conditions a fire could not have happened because the concentration of the gasoline vapour was too rich. But in this case the manhole had been left open inadvertently following some maintenance work on the stuffing box. Thus, air was able to blow into the vessel to form an explosible mixture with the vapour. The plant managers suspected that an electrostatic discharge was the cause of the fire as a result of feeding the sodium phenolate into the vessel via the PE tube. They asked us to confirm this view. It was generally accepted that the passage of the product through the tube had caused the latter to become charged; but the question was which type of gas discharge had taken place, if any?

Propagating brush discharges were not to be expected because the speed at which the product moved through the tube was insufficient to produce the required charge density (see 3.3.5). The charge density might have been high enough for a brush discharge, but because the tube was inside the vessel there were no grounded electrodes to initiate such a discharge. What alerted us was the hasty maintenance of the stuffing box which we suspected had not been done properly. We asked the management for someone to dismantle the stuffing box so that we could look for tarnishing colours which would indicate overheating. In fact, bluish colours were visible on the agitator's shaft in the region of the stuffing box packing. Doubtless, the latter had been adjusted too tightly. A specialist stated that a temperature of about 400°C had been reached at that place. As the ignition temperature of the gasoline is 260°C it was considered that it had been ignited and that the hot surface was the source of the ignition.

*Suggested safety measures*

- In the presence of combustible atmospheres, especially those with low ignition temperatures, slide-ring sealings are preferred to stuffing box sealings.

## 9.5 Fire inside a filter casing at the top of a silo

The accident occurred in the drying and storage area of a processing plant for plastic powder. The wet powder was first conveyed into a steam heated flash-dryer and then into a 200 $m^3$ silo. At the centre of the top of the silo was mounted a filter casing containing a number of filter hoses used for cleaning the dust laden air. On being told by the safety engineer that there had been an explosion and a fire in the filter casing we became curious to know how it had happened. Fortunately, the fire did not extend into the silo because it was almost full of product. Thus, only the surface of the settled product was scorched. As an aid to understanding our interest in this case it will be useful to describe the filter system.

By referring to Fig. 9.3 it can be seen that the hoses are suspended by hangers from the ceiling of the filter casing with their lower ends fitted into holes in the cover of the silo. The filters are cleaned by a vibrating mechanism (not shown). Each filter hose is itself supported by several steel rings sewn into the filter fabric which keep the filter in shape. Anyone familiar with electrostatics would be suspicious of these rings as they can become sources of spark discharges.

As this type of dust filter is in common use in industry because of its simplicity, adaptability and ease of handling, e.g. when changing hoses, we had already examined the possibility of incendiary discharges from the rings. Because of the insulating properties of the filter fabric the rings were, effectively, insulated from ground. Tests showed that their capacitances to ground were in the region of 5–10 pF. Taking the higher value and using the energy equation this means that the potential on a supporting ring must reach at

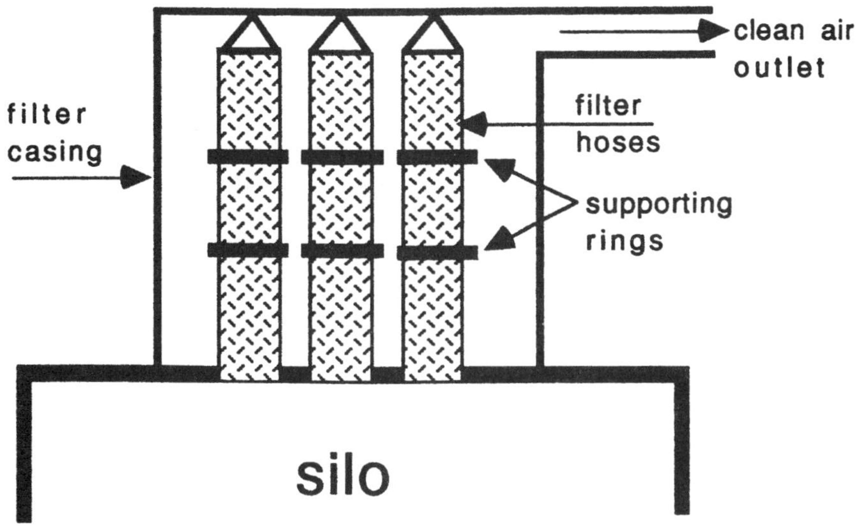


**Figure 9.3** Filter on top of a silo

least 45 kV to produce a spark discharge with an energy of 10 mJ. So, tests were done to find the maximum potential on the rings under proper running conditions. This turned out to be only 15 kV, which was rather disappointing and was found to be the result of local corona discharges which were limiting the level of potential on the rings. One might ask why no instructions were given to ensure that the rings were grounded? In fact, tests on another unit were done to electrically interconnect the rings by the use of metal ribbons sewn into the fabric. The ribbons were connected by cables to ground. However, these measures, which had been taken during the mounting of the hoses and each time the hoses were changed, proved to be very unreliable. Because of the vibration of the filter hoses during cleaning the integrity of the connections between the rings and ground could not be guaranteed. Further, several isolated but interconnected rings would present a greater danger than a single ungrounded ring because of their higher electrical capacitance. It therefore seemed appropriate to suggest that the rings remain isolated from ground and from each other.

After having shown that it was impossible to ground the rings reliably we felt very uneasy about approving the continued use of the filters. After all a fire had occurred! The only other possible cause of the fire that occurred to us was that two supporting rings might have made contact when a filter hose accidentally fell from the ceiling. This could have doubled the energy of a possible spark discharge from the rings. However, as this seemed to be so unlikely we stuck to our view that the rings should not be grounded. Instead, we suggested we examine closely the thermal stability of the product.

At first glance there was nothing to be suspicious about. The maximum temperature of the heated air in the conveyer flash-dryer was 150°C and the ignition temperature of the plastic powder was 470°C. However, a long time thermal stability test showed the following results:

- At 125°C – no reaction after 100 h.
- At 135°C – an exothermic reaction with a rise of temperature up to 300°C after 30 h.
- At 145°C – a rise of temperature up to 350°C after 16 h with the sample burned to a cinder.

By using these results a possible cause of ignition was deduced as follows. The product, on being introduced into the flash-dryer, might have settled to a level near to the incoming hot air. There it could have remained for a long period of time exposed to a powerful draft of air at a temperature of 150°C. A self-igniting reaction could have occurred with a corresponding rise in temperature of the product. Very hot cinders of the product could have been entrained into the airstream, taken to the dryer and on to the silo where the dust was ignited.

*Suggested safety measures*

- Long-term thermal stability tests should be done during all drying procedures.

## 9.6 Fire in a solvent cleaning area

The solvent used in a processing plant was acetone which had to be cleaned for re-use. Some storage vessels which could be interconnected with conducting rubber tubes had been installed in an area where, subsequently, a fire occurred. Shortly before the fire a worker noticed that one of the tubes was leaking. Acetone was running out and dripping through the egg-crate decking onto the floor beneath. He immediately switched off the pump, closed the valves and went for a replacement tube to the receptacle in which they were kept. But on that day, of all days, the tubes had been removed for routine inspection without the worker being informed. The only one left was a transparent PVC tube with a diameter of about 30 mm. The worker took it and was glad to find that it was the right size for the job. He connected the storage vessels by sliding the PVC tube onto the appropriate nozzles, switched on the pump and opened the valves thereby causing the acetone to flow again. Just a few seconds later a colleague working underneath shouted 'fire'. Being quick-witted, the worker switched off the pump and ran away before the flames could catch him through the egg-crate decking!

As the storage vessels were of steel the fire was limited to the spilled acetone which had come from an insecure rubber tube. The workers put out the fire themselves. Having recovered from the shock the persons involved wondered what could have been the cause of the ignition.

Every so often a chain of unfortunate coincidences can lead to such accidents. The worker was not one of the routine staff, as he had stood in for a sick colleague. Consequently, he was not informed about the inspection of the rubber tubes that day. On the other hand, the man whose job it was to inspect the tubes and replace the damaged ones was not instructed to leave some in the receptacle in case of an emergency. He took them all to the workshop. The workers in this plant were instructed to use only conducting rubber tubes with solvents because of possible electrostatic discharges. They were aware that transparent PVC could be dangerous from the electrostatic point of view. Even so they sometimes used them without permission during maintenance operations when checking the flow of the solvent. As the temporary worker was not informed about the danger of static discharges he could not be blamed for using the PVC tube to get on with his work. To the safety people the matter was quite clear. It must have been charge on the PVC tube which led to the ignition of the acetone vapour/air mixture. They admitted ungracefully that there had been a lack of liaison and organization. New operating instructions would be put in place immediately and the statics expert would be informed that a static discharge was again the cause of an accident.

To us this explanation seemed just too easy and we felt very suspicious. Of course, it could not be denied that transparent PVC, being a highly chargeable material, presents a possible electrostatic hazard. The tube could become charged simply by handling it, let alone passing acetone through it. Such charging could lead to a brush discharge capable of igniting an optimum mixture of acetone vapour and air. However, to establish that a discharge was the cause of the accident it was necessary to do some more research. First, the whole scenario was reviewed. As everything was found to be as reported, we casually looked at the various items of electrical equipment including lamps, wiring, switches and the pump motor which was mounted on the lower floor under the egg-crate decking. The entire equipment was licensed for use in Zone 1 and appeared to be perfect. The whole area was blackened by the fire but on closer inspection the pump motor was found to be blacker than the rest. The electrician was asked to open the terminal box of the motor. In Fig. 9.4 is shown what came to light. The normally tightly sealed box was blackened inside, obviously by the fire. It was a 3-phase motor connected in star circuit, as is usual, but the holding nut on the right-hand side of the star delta bridge was missing and the bolt and brass bridge piece were melted in places. In fact the missing nut was found lying at the bottom of the box; obviously it had become loose because of vibration. We were puzzled to find a hole tapped in

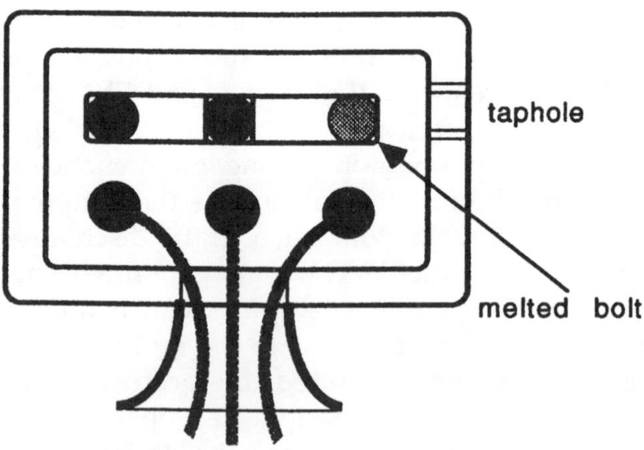


**Figure 9.4** Terminal box (open)

the side of the terminal box and questioned its use because it is agreed that boxes for explosion-proof motors have to be hermetically sealed. On inspecting the hole carbon black was found inside it. At this instant we understood the cause of the problem.

Acetone had dripped from the upper floor via the egg-crate decking onto the terminal box of the pump motor. The solvent was able to seep through the hole in the side of the box. When the pump was switched on again, after attaching the PVC tube, electrical sparks had occurred at the nutless screw on the star delta bridge. These sparks had ignited the acetone vapour in the box and the flame had passed through the hole thereby setting fire to the acetone surrounding the motor.

The question which still intrigued us was why the hole was drilled in a box which was meant to be sealed. On further investigation it was found that several years ago an electrical engineer had issued an instruction for extra ground connections to be made to all the metallic casings of electrical equipment to ensure their equipotential. In order to bolt this connection to the casing a taphole had to be drilled. His intention had been to provide greater safety against the possibility of shocks and sparks. Several years later his successor was able to show how much money could be saved if these extra groundings were no longer installed. Instead, he suggested that the normal grounding be improved. It was evident that a repair had been made to the pump which required the pump motor to be dismounted. At the time the pump was reassembled the instruction to fix no extra grounding was already in force and so it was left out. Neither the electrician nor anybody else noticed that there was a hole still in the box!

*Suggested safety measures*

- Doing more to improve safety can make things worse!
- Too many cooks spoil the broth!

## 9.7 Literature

(1) Bartknecht, W. (1989). Zundwirksamkeit von mechanisch erzeugten Funken und heissen Oberflachen in Staub/Luft- und Brenngas/Luft-Gemischen. VDI Fortschrittberichte No. 180. VDI Verlag, Dusseldorf.

# Chapter 10
# Shocks to people caused by static electricity

## 10.1 Metal crates containing plastic bottles

Polyethylene bottles of about 50 mm diameter and 150 mm in height were being filled manually with strips of paper coated with plastic. After a cap was screwed onto a bottle the latter was placed into a metal crate which could hold 100 bottles. The crates were moving slowly on a conveyer belt passed several workers who filled them with bottles. As clean room conditions were required in the area all employees handling the strips had to wear latex gloves. The relative humidity of the room needed to be kept at a low level. A worker at the end of the conveyor belt, whose job it was to remove the crates to a trolley, was constantly troubled with electric shocks on touching the filled crates. As the bottles were already closed she did not need to wear gloves. However, on the occasions when she was wearing gloves she did not feel a shock every time she handled a crate, but when a shock was felt it was more severe than any received by her bare hands.

As one would imagine, the job of transferring the crates to the trolley was the most unpopular one in the plant area and it soon became necessary to remedy the problem which, obviously, arose from static charges. Tests on the resistances to ground of the crates, via the conveyor belt, gave values of about 50 TΩ. The bottles were found to be negatively charged after being handled and the induction effect from the charge on 100 bottles was to produce potentials on the crates in the region of 7–9 kV. While the shocks from the crates were uncomfortable there was no doubt that they presented no real danger to the workers.

The infrequency of shocks to the operator and their greater severity when she was wearing gloves implied that the breakdown potential of the latex gloves corresponded with the highest potentials reached on the crates.

*Remarks*

The often practised measure of grounding people by, for example, the use of conducting footwear to avoid electrostatic shocks would not have been helpful in this case because the shocks arose from charge on the crates, not on the people. Therefore, it was the charge on the crates which needed to be dissipated to ground. To achieve this might be thought to be a simple matter but, in fact, in this case it proved to be very difficult. Sliding metal contacts were elaborate to apply and were too delicate in use to be practical. A conducting conveyor belt would be costly and would rely on carbon black for its conducting properties. As the workplace was a pharmaceutical cleanroom such a belt would not be permitted anyway in view of the possible abrasion of carbon particles and their deposit onto the product. Neither was an active corona neutralizer (see 3.3.1.1.1) considered suitable because of its steady generation of ozone.

The only remaining solution was to replace the metal crates with plastic ones. Although this caused us to feel uneasy about the possibility of super brush discharges (see 3.3.3 and 10.5) we heard nothing more from the company. In our experience, when your advice works well you are unlikely to receive any further response from the customer.

## 10.2 Cutting PE foam

PE foam is produced continuously in large blocks of cross-section of about 2 $m^2$ and is then cut to a desired size by means of a bandsaw. People operating a bandsaw in a plant where this type of foam was manufactured complained of electric shocks on touching any metal surface in the vicinity of the machine. It was clear that static charges produced during the cutting of the highly insulating foam was the cause of the problem.

The facts of the case were quite clear. The electric field from the charged foam was causing the operators to become charged by induction. To avoid the unpleasant consequences of this the induced charge needed to be quickly dissipated to ground. The remedy would normally have been to install conducting flooring and ensure that any workers near to the machine wore conducting footwear. Unfortunately, the management considered such action to be impractical because of the high cost. It would have meant replacing the insulating PVC flooring with a conducting material and equipping the relatively high number of people going to and from the machine, to take away cut foam, with conducting footwear. To resolve this problem a less expensive solution had to be found. In avoiding electrostatic nuisance it is, in general, sufficient merely to reduce the amount of charge to a level at which it ceases to be a problem without eliminating it altogether. This can be achieved simply by the use of passive neutralizers (see 3.3.1.1.2). These are easily constructed and in the present case it was enough to stretch a grounded metal wire of about 0.3 mm diameter across and above the surface of the foam at the place where it emerged from the cutting machine.

There were no further complaints of static shocks from the workers.

*Remarks*

This is a good example of how effective simple passive ionizers can be in overcoming static problems.

## 10.3 Carbon dioxide fire extinguisher

Workers at an electrolysis plant had at times to cope with the problem of small fires which were easily put out by the use of a carbon dioxide fire extinguisher. However, each time they used it they were troubled with severe electric shocks on touching any nearby large or grounded metal surfaces. As they had great respect for the heavy currents used in the electrolysis plant, which could be a danger to life, they took the problem of the shocks very seriously. For protection against mains current the workers were equipped with electrically insulating footwear and the floor was covered with a layer of insulating material. When the fire brigade informed us about the problem of shocks we immediately suspected static charge on the workers as the cause because they were insulated from ground. On checking the resistances to ground of several workers, values in the range 1.0–10 TΩ were measured. Assuming an average electrical capacitance for those tested of 150 pF, the corresponding charge decay time would be between 150 and 1500 s. But the question remaining was how could charges which were sufficient to cause severe shocks have accumulated on the workers while they were using the extinguishers? Obviously, the charging of the workers must have occurred as the carbon dioxide was expelled from the extinguisher. But this is at odds with the statement made in 2.2.2 that flowing gases will not of themselves give rise to static charges. However, on examining an extinguisher more closely during use it was observed that a kind of snow was expelled from it which could only have been particles of solid carbon dioxide produced by the marked cooling of the gas as it suddenly expanded. Because of these solid particles there was the possibility that the extinguisher itself could become charged. To examine this the following tests were done. A 5 kg carbon dioxide extinguisher was held by a worker who was insulated from ground by his standing on a sheet of PTFE. A static voltmeter with a range of 0–25 kV was connected to the extinguisher. Within 2 s of the extinguisher being set off the voltmeter indicated full-scale deflection and the test was stopped. When the worker holding the extinguisher touched a grounded metal surface a static discharge occurred and the worker felt a painful shock.

As a means of overcoming the problem, grounding the worker was not permissible because of the need for safety against mains current. The remedy was to replace the metal nozzle of the extinguisher with a plastic one in order to restrict the flow of the charge, separated at the nozzle, to the body of the extinguisher. In a similar test to that described above an extinguisher fitted with a plastic nozzle showed a potential of less than 3 kV after 2 s and about 10 kV after 20 s when all of the gas had been released. As the small fires in the plant could be put out by short blasts (less than 3 s) of gas, the problem was solved.

*Remarks*
We ask ourselves if the use of carbon dioxide extinguishers can present a danger from electrostatic discharges. There is no doubt that the spark discharges which occur when using these extinguishers can ignite flammable gases and vapours and, possibly, certain combustible dusts. As fire extinguishers are, of course, used in the presence of fires an additional ignition source becomes irrelevant. However, should an extinguisher be set off accidentally in an explosion-endangered area, it could cause an ignition. Such cases have been reported in the past.

## 10.4 Propagating brush discharge at a PP expansion pipe

A 10 $m^3$ agitator vessel which was half filled with an aqueous suspension of an intermediate organic product was kept under an air pressure of 6 bar (0.6 MPa). In order to release the pressure a worker opened a valve which was connected to a 125 mm diameter PP exhaust pipe. After the pressurized air had been passing for about 20 s the worker, while holding on to a metal part of the rig with one hand, touched the surface of the PP pipe with the other. He immediately felt a severe electrostatic shock which threw him from the platform on which he was standing onto the floor. He sustained injuries which required that he be taken to hospital straight away.

In attempting to discover the cause of the accident the first approach was to the electrician who was asked if it could have been a stray mains current that injured the worker. His tests showed that there were no electrical potentials on the metal surface held by the worker nor on the platform.

The plant managers, being at a complete loss as to the cause of the accident, asked the advice of the 'electrostatics bloke'! After all, how could clean air from an aqueous suspension cause static electrification? A full-scale test was begun to find out what was going on. An electrostatic fieldmeter was installed at a distance of 10 mm from the pipe, near to the place where the worker had supposedly touched the pipe. The vessel was half filled with water and an air pressure of 5.5 bar (0.55 MPa) was applied. Then the valve to the exhaust pipe was opened. We were surprised by the noise of the flowing air and by the sharp crack which followed. The fieldmeter indicated a zero reading but the pen of the recorder to which it was connected was stuck at full-scale deflection! Wondering what had happened we waited until the air had been released and then pressed the test button on the fieldmeter. The meter was broken.

Thus, the sequence of events had been first, that a man was badly injured after receiving a shock, then, after hearing a sharp crack during our tests, the fieldmeter was found to be broken. Was it possible that we were seeing the effects of propagating brush discharges? We carried out another test in which a grounded metal sphere of 100 mm diameter was placed at a distance of about 10 mm from the exhaust pipe. A few seconds after the valve was opened a sharp crack was heard again and a luminous spark discharge was observed between the metal sphere and the pipe. The spark extended in a branch-like fashion over the surface of about a 1 m length of pipe. This was clearly a propagating brush discharge.

On examining the pipe with a magnifying glass several crater-like markings could be seen. These were typical of traces left by propagating brush discharges. But how could the high charge density on the pipe necessary for such a discharge have occurred? We dismantled the pipe and looked inside it. Instead of finding the inside surface to be dry, as was expected, it was partially wetted. This indicated that not only air but also water aerosols had been transported through the pipe. Water is itself conducting but if when tiny droplets are deposited on an insulating surface no continuous layer is formed, charge separation can occur leaving the surface in a charged condition (see 8.29).

To avoid the continuance of further propagating brush discharges the PP pipe was replaced with one of stainless steel.

*Remarks*

As the electric field associated with a propagating brush discharge resides mainly within the dielectric material on which the charge is stored, the field outside the dielectric is relatively weak. It follows that in checking for the possibility of a propagating brush discharge the use of a fieldmeter is unsuitable.

## 10.5 Loading parcels into a postal van

A Post Office worker was busy throwing parcels into a postal van. The parcels contained mail order catalogues which were sealed in polyethylene foil. While working he realized that the way he was stacking them would not allow all the parcels to fit into the van. So he climbed up to the loading area to stack the parcels a little higher (Fig. 10.1). According to his statement later, he was holding on to a reel with one hand while reaching for the pile of parcels with the other when he was startled by a painful electric shock.

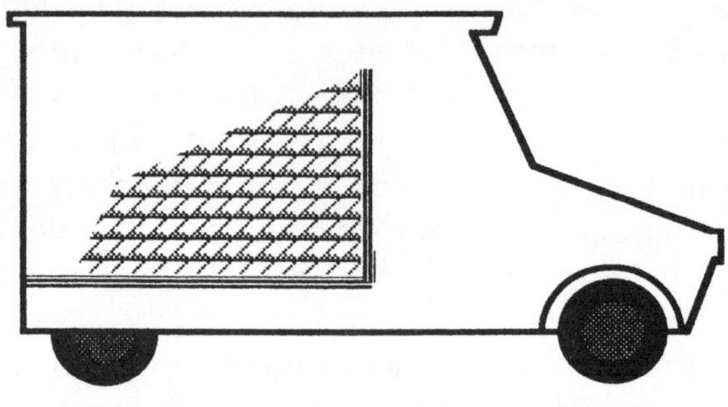

**Figure 10.1** Post Office parcel van

What had happened?

As the van had only a 12 V system and the engine was not running at the time, any accident caused by the van's electric circuit could be ruled out. However, an electrostatic discharge was soon suspected because of the plastic sealed catalogues. When thinking about the different types of gas discharge it seemed that it might be a rare incident of a super brush discharge (see 3.3.3) having taken place.

The sixty-four thousand dollar question that had to be asked was could such a discharge be created during the loading of the catalogues? Well, the postman took the parcels from a moving rubber belt where electrostatic charge separation took place between the belt and the parcel covers. As the same materials were making contact and separating from the belt all the time, the parcels would each have on them the same polarity of charge. On throwing the parcels into the van the repelling coulomb forces of the unipolar charges would be overcome by the much higher gravitational forces on the catalogues, resulting in an increase in the density of the charge on the parcels. The postman, being electrically conducting, was grounded via the reel and when his hand approached the parcels a painful super brush discharge was initiated.

## 10.6 Static in motor vehicles

In the context of electrostatic nuisance the everyday use of motor vehicles gives rise to persistent and often uncomfortable shocks to the people using them. Most sufferers complain of having received shocks when after leaving their vehicle they touched the car or garage door. However, some people do not appear to be susceptible to such shocks. An analysis of the habits of the latter has shown that they usually wear shoes with leather soles, or they have an older car.

A number of ideas are being discussed by those concerned about how the problem of static in vehicles can generally be avoided, e.g. grounding the vehicle by means of conducting rubber strips which make continuous contact with the road, or applying antistatic treatments to the seat covers, etc. These measures have each proved to be more or less inefficient. To discover what is really happening it has been necessary to question a number of vehicle users about their experiences. The results have shown that, in general, there is not a problem with old fashioned vehicles, no matter what type of shoe sole the users wear. However, with modern vehicles users do experience problems when they are wearing, for example, shoes with caoutchouc soles or any other type of sole which is a good electrical insulator.

This leads to argument and misunderstanding between drivers and passengers when those wearing leather soled shoes have no problem while those wearing shoes with insulating soles suffer shocks. Such a conflict can only be resolved by making appropriate electrostatic measurements under realistic conditions.

The measured potentials on persons wearing shoes with caoutchouc soles as they stepped out of different types of contemporary car were, surprisingly, found to be as high as 12 kV! However, when the shoes were changed for leather soled ones scarcely any potential was observed. It is well known that leather soles are electrically dissipative. Tests on the same people leaving an old fashioned car showed potentials which could hardly be detected no matter what type of shoe sole was worn.

These results lead us to ask what are the differences between modern and old fashioned cars which cause them to have such different electrostatic behaviour? The seat covers (wool, synthetic, leather) could be checked for their electrostatic properties, but this would not tell us a lot as their charging behaviour depends on the nature of the materials (jackets, trousers, skirts, etc.) which make contact with them.

On examining an old fashioned car it was found that many metal parts, e.g. door handles, window cranks and dashboards were in electrical contact with the body of the car. However, when attempting a do-it-yourself electrostatic check-up on a modern car we found that almost everything was made of insulating plastic material or was covered with plastic. Bearing in mind that a person alighting from a car usually has hold of the door handle we were able to make certain deductions. When a person leaving a car slides out of the car seat there is a separation of charge at the interface between the seat cover and the person's clothing. The charge on the clothing instantly induces a similar charge onto the body of the person. Similarly, the charge on the seat cover (of opposite polarity to that on the clothing) induces the same charge onto the body of the car. As there is an electrical contact between the person and the body of the car, through a conducting door handle, the opposite charges on the person and the car become neutralized without causing a shock to the person. However, if there is no electrical connection between a person wearing insulating shoes and the body of the car because, say, of a plastic door handle, the induced charge and the corresponding potential remain on the body. When the person touches a metal part of the car, such as the door as it is slammed shut, a spark discharge occurs resulting in a shock.

Some people are curious to know what happens to the charge induced onto the car by that on the car seat cover. The answer is that because of the low electrical resistance of the tyres the charge is dissipated to earth in a fairly short time, depending on the conducting nature of the road surface. Despite this there are people who are adamant that it is electrostatic charge on the car which is the cause of the shocks. Yet, they may readily be convinced of the error in their views by simply removing their shoes and stepping out of their car in their socks (or bare feet). They will not feel a shock.

So what is the solution to this problem for today's vehicle manufacturers? The answer is to provide a conducting door handle, or some other conducting part, which is connected to the body of the vehicle. Simple though this remedy is, the manufacturers seem to disregard it!

We overcome the problem by pushing on the glass window when slamming the door!

# Index